COASTAL AGRICULTURE AND CLIMATE CHANGE

About the Editors

Professor Dr. M. Prakash born in 1966, obtained his B. Sc (Ag.) and M. Sc (Ag.) from Tamil Nadu Agricultural University, Coimbatore, India in 1988 and 1990 respectively. He did his Ph. D from Annamalai University. He was awarded with National Merit Scholarship, Government of India during 1981-1988 and ICAR Junior Fellowship during 1988-1990. His expertise in abiotic stress tolerance is well recognised by invitations to serve in the Editorial Board / Committees of several academic journals, awards and Fellowships like Crop Research Award in 2004, Best Researcher Award in Annamalai University in 2010, and Fellowship like FELLOW of Indian Society of Plant Physiology, New Delhi in 2015 (FISPP) and FELLOW of National Academy of Biological Sciences, Chennai, 2015 (FNABS). Professor Dr. M. Prakash is having around 26 years of teaching and research experience. He has guided 8 Ph.D. scholars and 22 PG Scholars. His credentials include publications of 85 refereed research papers, seven books published by National publishers, and one book chapter published in SESAME by globally reputed CRC Press. He organized 6 National Seminars and completed five research projects. He has earned the status of Departmental Research Support with Special Assistance Programme of University Grants Commission (UGC-SAP- DRS) to the Department with financial assistance of Rs. 50 lakhs at Phase I(2009-2014) and Rs. 1.025 Crore at phase II (2015-2020) level.

Dr.S.Arivudainambi, Professor of Entomology, Faculty of Agriculture, Annamalai University. Has 25 years of teaching experience. He was awarded with the Best research paper award during 2011 at Annamalai University, Best faculty award during 2014 by ETC and Best researcher award during 2018 at Annamalai University. He specializes in phyto insecticides and insecticide toxicology. He had good experience in mass production and evaluation of insect viruses. He discovered a novel insecticidal principle from Cleistanthus collinus, a wild shrub belonging to family Euphorbiaceae, and it is under the process of formulation. He has authored more than 100 articles and three books. He served as expert member to guide and evaluate research projects funded by ETC- COMPAS at Netherlands.

Dr. S. Rameshkumar is working as Professor of Horticulture in Department of Horticulture, Faculty of Agriculture, Annamalai University. His area of specialization is floriculture, post harvest technology and environmental horticulture. He has rich experience of Landscape designing and maintenance in Annamalai University for more than 12 years. He has 20 years of teaching and research experience. He has guided three Ph.D. scholars and 15 PG scholars. He has published around 25 research articles in reputed journals. He has organized a National Symposium on Horticulture in the Vanguard of Climate Change and Urban Environment during Feb, 2019. He is conferred with Fellow of National Gladiolus Trust. He has served as associating scientist in ICAR-NAIP project and DST-BIRAC project in Faculty of Agriculture.

Dr. S. Babu, is teaching soil fertility and weed management at Faculty of Agriculture, Annamalai University for well over a decade. He took UG, PG and Ph.D.degrees from Annamalai University. After graduation he worked as Project Officer in Monsanto India Ltd., Program co-ordinator in All India Radio, Puducherry and Training Associate (Agronomy) in krishiVigyan Kendra, Krishnagiri. Then he joined as lecturer in 2004 in the department of Agronomy, Faculty of agriculture.He has participated in number of national and international conferences, seminars and workshops. He has 15 years of teaching and 18 years of research experiences and published 21 research papers in various national and international scientific journals. He also published more than 27 popular articles in various newspapers and magazines. His specialization is Soil fertility & Nutrient and Weed management. He has been serving as a state level resource person and gave more than 87 invited lectures to NGO, state Agricultural Department in Tamil Nadu & Puducherry, KVK and Agricultural colleges.

COASTAL AGRICULTURE AND CLIMATE CHANGE

Editors

M. Prakash
Professor
Annamalai University, Annamalai Nagar
Chidambaram, Tamil Nadu-608002

S. Arivudainambi
Professor of Entomology
Faculty of Agriculture
Annamalai University, Annamalai Nagar
Chidambaram, Tamil Nadu-608002

S. Rameshkumar
Professor of Horticulture
Department of Horticulture
Annamalai University, Annamalai Nagar
Chidambaram, Tamil Nadu-608002

S. Babu
Faculty of Agriculture
Annamalai University, Annamalai Nagar
Chidambaram, Tamil Nadu-608002

NEW INDIA PUBLISHING AGENCY
New Delhi-110 034

NEW INDIA PUBLISHING AGENCY
101, Vikas Surya Plaza, CU Block, LSC Market
Pitam Pura, New Delhi – 110 034, India
Email: info@nipabooks.com
Web: www.nipabooks.com

For customer assistance, please contact
Phone: +91-11-27 34 17 17 Fax:+91-11-27 34 16 16
E-Mail: feedbacks@nipabooks.com

ISBN : 978-93-901750-01-7

Composed and Designed by NIPA.

Preface

Changing climate in micro and macro level attribute to influence and bring changes in productivity and production trend of every industry including agriculture and its ecosystem. The influence of climate change and its environmental impact on coastal ecosystem greatly influence the life and livelihood of agrarian communities and people involved in allied industries. The vast coastline of India covering the major areas of farm lands in 66 districts are prone to salinity and flash flood problems making the coastal agriculture challenging. It is need of the day to highlight and focus the technologies pertaining to coastal agriculture to augment the productivity of the crops in coastal farming systems.

Sustainable environmental management strategies in coastal ecosystem need to be addressed to save the coastal areas of India for future generation. Hence, this book has been authored with objectives of providing understanding on the environmental problems, features of coastal ecosystem, advanced strategies for environmental protection, mitigation strategies for environmental problems in coastal areas and augmenting agriculture income from coastal areas. In this context, publication of this book on

"Coastal Agriculture and Climate Change" is most appropriate and need of the hour. We are confident that this book will certainly be useful for the faculty and researchers in understanding problems and prospects of Coastal Agriculture in the present scenario of Climate change.

Editors

Contents

1

Coastal Agriculture Status and Strategies

M. Prakash, G. Sathiyanarayanan, B. Sunilkumar and K.R. Saravanan
Department of Genetics and Plant Breeding, Faculty of Agriculture
Annamalai University, Annamalai Nagar-608 002, Tamil Nadu, India

Coastal areas are defined as the interface or transition areas between land and sea, including large inland lakes. Coastal areas are diverse in function and form, dynamic and do not lend themselves well to definition by strict spatial boundaries.

Economic importance of coastal areas

Many of the world's major cities are located in coastal areas, and a large portion of economic activities are concentrated in these cities. The coastal zone is an area of convergence of activities in urban centres, such as shipping in major ports, and wastes generated from domestic sources and by major industrial facilities. Many of the world's most productive agricultural areas are located in river deltas and coastal plains, which contribute for national economic growth.

Opportunities and Constraints for Agriculture in Coastal Areas

A. Opportunities

There are three types of opportunities, which are being potentially important in the development of agriculture in coastal areas.

1. **Opportunities dependent on natural resources**

   In general, coastal areas offer very favourable conditions for agriculture, because of alluvial accumulation plains. Such areas generally have deep, relatively flat, fertile soils and also benefit from a substantial supply of water. Most of the coastal areas have a milder and more humid climate than the interior as a result of the moderating influence of the sea, especially where favourable sea currents occur, which is favourable for growth of a particular crop or crops which are not commonly grown elsewhere in the country.

2. **Opportunities arising from location**

   Since these areas are located near the coast, the transport costs for its produce are less when compared to inland agriculture. Coastal roads give easier access to markets for agricultural products, and also facilitate input supplies. Even in the absence of good roads, produce can be shipped to markets by river by boat along the coast.

3. **Derived or secondary opportunities**

   Coastal industries and their development may lead to population growth, thus increasing the demand for food. When parts of the population are becoming wealthier, there may also be increasing demand for higher-quality foodstuffs, such as fruit, vegetables, meat and dairy products. Growth in the coastal industrial growth may lead to increased demand for agricultural raw materials.

B. Constraints

Constraints specifically affecting agriculture and agricultural development in coastal areas, arise principally from:

- The direct effects of the proximity of the sea;
- The location of coastal agriculture at the downstream end of river flows;
- The limited space for expansion or relocation and the limited resources in many coastal areas.

1. **Proximity of the sea.**

   Due to close proximity of sea, low-lying agricultural lands may be susceptible to flooding as a result of coastal erosion, subsidence or a rise in sea levels. Higher air humidity in coastal areas may be favourable to the occurrence and propagation of certain plant diseases and pests which will affect crop growth. Penetration of sea water in the inland increases river water salinity which may complicate its safe use for irrigation.

2. **Upstream effects.**

   Upstream dams and irrigation schemes not only deprive coastal areas of water for irrigation but also affect the quality of water available to coastal agriculture by upstream discharges of industrial and urban effluents and drainage of chemicals and salts from various sources.

3. **Space and resource constraints.**

   There is limited scope for coastal areas for further extension. Opportunities for further expansion or relocation of agricultural activities are therefore limited, especially when the coast is bounded by mountains. Because of space constraints, agricultural activities such as grain crops and livestock grazing may be marginalized and replaced by non- agricultural activities.

Coastal agriculture is largely affected by salinity and submergence.

A. Salinity

Over 20 million hectares of land suited to rice production in Asia are currently either underexploited or unexploited because of excess salt and other related soil problems. In India and Bangladesh alone, productivity of more than 7 million hectares of rice land is adversely affected by salt stress.

Rice is suitable for rehabilitating these soils because of its ability to grow under flooding and its high potential for genetic improvement. Rice productivity in salt-affected areas is very low, < 1.5 t/ha, but can reasonably be raised by at least 2 t/ha, providing food for more than 10 million of the poorest people living off these lands.

The problem of coastal soil salinity in India encompasses

1. Coastal saline soils situated in humid and sub-humid areas
2. Coastal saline soils of arid areas and
3. Coastal acid saline soils.

Salt affected soils occur within a narrow strip of land adjacent to the coast and up to 50 km wide. These areas generally have an elevation of less than 10 m above mean sea level and include the low-lying land of river deltas, lacustrine fringes, lagoons, coastal marshes, and narrow coastal plain or terraces along the creeks.

There are two approaches for coastal saline management viz.,

1. Varietal approach
2. Crop & Soil management approach.

1. Varietal approach

Salt-tolerant rice varieties also offer great potential to grow rice in marginal lands, which are usually left fallow particularly during the dry season because of high salinity.

Development of varieties tolerant to salinity are preferable because the management practices are to be follower for every season whereas once the varieties are developed, they can be for years together. Saline tolerant varieties are preferable because

1. Salt tolerant varieties of crops require less chemical amendments,
2. Varietal approach is simple, cheap and eco-friendly and
3. Suitable for poor resource farmers due to low cost

Salt tolerant varieties

- **Rice:** CSR-49, CSR 36, CSR 30 (basmati type), CSR 27, CSR 23, CSR 13 and CSR 10.
- **Rice variety for coastal regions:** Butnath (CSRC(S) 5-2-2-5) and Sumati-CSRC-CSRC(S) 2-1-7.
- **Wheat:** KRL 213, KRL 210, KRL 19 and KRL 1-4.
- **Indian Mustard:** CS 58, CS 56, CS 54 and CS 52.
- **Chick pea (gram):** Karnal Chana 1.
- **Dhaincha (Sesbania):** CSD 137 and CSD-123.

Saline tolerant rice varieties released for coastal areas

Name	Country	Year
BINA Dhan 8	Bangladesh (coastal saline soils)	2010
BRRI Dhan 52	Bangladesh (coastal saline soils)	2010
BRRI Dhan 53	Bangladesh (coastal saline soils)	2010
BRRI Dhan 55	Bangladesh (coastal saline soils)	2011
CR Dhan 405	India (Orissa & West Bengal)	2012
CR Dhan 406	India	2012
BRRI Dhan 61	Bangladesh	2013

2. Soil & Crop Management approach

Coastal areas are different from inland areas. Salinity problems in coastal soils is caused during the process of their formation under marine influence and subsequently due to periodical inundation with tidal water and in case of lowlands having proximity to the sea due to the high water-table with high concentration of salts in it. In coastal and delta regions of major rivers of the world the soils are rich in salts due to the presence of saline ground water-table at shallow depth.

(a) **Scraping**: Removing the salts that have accumulated on the soil surface by mechanical means.

(b) **Flushing**: Washing away the surface accumulated salts by flushing water over the surface is sometimes used to desalinize soils having surface salt crusts.

(c) **Leaching**: Removing salts from the root zone of soils. Leaching is most often accomplished by ponding fresh water on the soil surface and allowing it to infiltrate.

(d) **Mulching**: Mulching with crop residue, such as straw, reduces evaporation from the soil surface which in turn reduces the upward movement of salts. Reduced evaporation also reduces the need to irrigate. Consequently fewer salts accumulate.

(e) **Deep Tillage**: Accumulation of salts closer to the surface is a typical feature of saline soils. Deep tillage would mix the salts present in the surface zone into a much larger volume of soil and hence reduce its concentration and impact. Many soils have an impervious hard pan which hinders in the salt leaching process. Under such circumstances "chiselling" would improve water infiltration and hence downward movement of salts.

(f) **Incorporation of Organic matter**: Incorporating crop residues or green-manure crops improves soil tilth, structure, and improves water infiltration which provides safeguard against adverse effects of salinity. In order to be effective, regular additions of organic matter (crop residue, manure, sludge, compost) must be made.

(g) Conservation farming practices to control soil salinity by following reducing summer fallow, using conservation tillage, adding organic matter to the soil and planting salt-tolerant crops (e.g., rapeseed and cabbage).

(h) Maintenance of satisfactory fertility levels, pH and structure of soils to encourage growth of high yielding crops.

(i) Maximization of soil surface cover, e.g. use of multiple crop species.

(j) Selection of crops *viz.*, use of deep-rooted plants to maximise water extract.

(k) Using crop rotation, efficient irrigation of crops, soil moisture monitoring and accurate determination of water requirements.

(l) **Goa Bio-1:** Bio-formulation for plant growth promotion of paddy under salt affectedsoils of coastal regions.

One of the promising approaches to improve the productivity of paddy in coastal areas is to improve the soil biological activity that would indirectly help in nutrient mineralization, alleviation of salinity stress, leading to better crop establishment and production. It can be done by augmenting the saline tolerant plant growth promoting microorganisms. ICAR-CARI had identified a promising salt tolerant plant growth promoting bacterium called Goa Bio-1.

Benefits of using Goa Bio-1

- Better nutrient mineralization,
- Alleviation of salinity stress
- Better crop establishment
- Improved plant growth parameters, yield and soil biological activity.
- This formulation is recommended in paddy grown under saline soils and vegetablecrops (Brinjal, tomato, chilli and cucumber), black pepper, fruit and plantation crop nurseries.

2. Goa Bio-2

Bio-formulation for plant health management of field, vegetable crops and black pepper in Coastal regions ICAR CCARI had identified a promising plant growth promoting bacterium with diswease supression ability.

Application of Goa Bio-2 improved plant growth parameters, plant health and yield. This formulation is recommended in vegetable crops (Brinjal, tomato, chilli and cucumber), black pepper, fruitand plantation crops and nursery plants.Soil borne disease incidence was significantly lower in Goa Bio-2 treated plants.

B. Flood tolerance

Coastal flooding is a sudden and abrupt inundation of a coastal environment caused by a short-term increase in water level due to storm surges created by storms like hurricanes and tropical cyclones and rising sea levels due to climate change. The magnitude and extend depends on the coastal topography and storm surge conditions.

Flood damages are expected to increase significantly over the 21st century as sea-level rise, more intense precipitation, and extreme weather events combined with socio economic developments due to increasing population. The average global coastal flood losses in the 136 largest coastal cities in the world have been estimated to be approximatively US$6 billion (Darren Lumbroso, 2017).

The Inter governmental Panel on Climate Change (IPCC) estimate global mean sea-level rise from 1990 to 2100 to be between nine and eighty eight centimetres (Nicholls, 2002). It is also predicted that with climate change there will be an increase in the intensity and frequency of storm events such as hurricanes (Dawson, 2009; Nicholls, 2007; Suarez, 2005). A rise in sea level alone threatens increased levels of flooding and permanent inundation of low-lying land as sea level simply may exceed the land elevation (Nicholls, 2002; Michael, 2007). This therefore indicates that coastal flooding associated with sea level rise will become a significant issue into the next 100 years especially as human populations continue to grow and occupy the coastal zone (Nicholls, 2007).

Floods annually reduce productivity in rice over 22 M.Ha in Asia and major area in Africa. Flood is a recurrent phenomenon in coastal Tamil Nadu, Andhra Pradesh, Odisha, West Bengal, Kerala and south Gujarat in India. Submergence stress is estimated to cause annual losses of up to $1 billion in Asia. Modern rice varieties are not adapted to these conditions and hence farmers suffer from either regular yield loss when they grow these varieties or low yield when they continue to cultivate local landraces.

Apart from improving drainage and other preventive measures, farmers can adopt flood tolerant varieties that can withstand inundation for an extended period and reduce the risk from flood damage.

Impact of flooding on Crops

- Most crops grown in coastal areas are intolerant of flooding.
- Flooding depletes soils of oxygen and increases disease infections and nitrogen losses.
- Weather conditions following flooding are important to plant survival. Cool, wet conditions favour disease development.
- Crop condition may not be apparent for several days after flooding. Check plant meristems to assess damage and recovery potential.
- Replanting should be a strictly economic decision that considers the yield potential of the reduced stand vs. that of a later-planted crop minus replant costs.

Coastal floods can be managed by (1) Varietal approach and by (2) Soil & Crop Management approach.

1. **Varietal approach**

   IRRI developed its first flood-tolerant rice variety in 2008, the IR 64 Sub 1, which came into fruition with the help of a traditional flood-tolerant variety in India. IRRI has discovered, in rice varieties in Orissa and Sri Lanka, a gene that they called Sub 1 gene, which enables rice to survive- and more importantly, recover-after flooding. IRRI has worked to add the Sub 1 gene to Swarna (a popular rice variety in India). Swarna with Sub 1 yields is twice that of Swarna without Sub 1, with about 15 days of submergence.

Submergence tolerant rice varieties released

Name	Country	Year
Narendra Dhan	India	2005
Varsha Dhan	India	2005
Narendra Maunk	India	2009
Narendra Jalpushp	India	2009
Narendra Narain	India	2009
Improved Suwarna	India	2009
Suwarna Sub 1	India	2009
BRRI dhan-51	Bangladesh	2010
Suwarna Sub 1	Nepal	2012
NDR 8011	UP, India	2016
IR 64-Sub 1	Other states, India	2016
CR 1009 SUB 1	Other states, India	2016
Cihereng-SUB 1	Nepal	2016
BR 78 (BRRI dhan-78)	Bangladesh	2016
Narendra Shishir	UP, India	2017
Narendra Neha	UP, India	2017
Amala, Swarnali	W. Bengal, India	2017
BRRI dhan-79	Bangladesh	2017
NDR 8015	UP, India	2018
NDR 9930111	UP, India	2018
NDR 9730018	UP, India	2018
NDR 8017	UP, India	2018
NDR 8017	UP, India	2018

Multi stress tolerant variety

Developing varieties tolerant to abiotic stress and biotic stresses separately requires lot of time, labour and also expensive. Again farmers have to rely on varieties tolerant to any one of the stresses and have to take management practices for other stresses. Hence in order to make the stress management simple and less expensive, the concept of development of multi stress tolerant varieties has been developed. Accordingly few multi stress tolerant varieties were developed.

Rc480 Popularly known as **GSR (Green Super Rice**) was developed by IRRI, Philippines. It is resistant to multiple abiotic stresses, such as drought, salinity, alkalinity, iron toxicity and has intermediate resistance to pests, such as Yellow Stem Borer (YSB) and Brown Plant Hoppers (BPH). It matures in 107 days after sowing (DAS) and gives maximum yield of 4.4 t/ha, despite lesser input requirement.

2. Soil and Crop Management approach.

The one way to prevent significant flooding of coastal areas now and in the future is by reducing global sea level rise. This could be minimised by further reducing greenhouse gas emissions. The technical measures for flood management include crop management techniques like foliar spray of KCL, growth regulators etc.

Technologies to manage this situation are:

1. Providing adequate drainage for draining excessive stagnating water around the root system.
2. Spraying of cycocel @ 500 ppm for arresting apical dominance and thereby promoting growth of laterals.
3. Foliar spray of 2% DAP + 1% KCl.
4. Spray of 40 ppm NAA for controlling excessive pre-mature fall of flowering/buds/young developing fruits and pods.
5. Spray of 0.5 ppm brassinolide for increasing photosynthetic activity.
6. Foliar spray of 100 ppm salicylic acid for increasing stem reserve utilization under high moisture stress.
7. Apart from the above, flood control measures also include, maintenance of natural dune systems, protection of coastal ecosystems and different flood proofing and accommodation activities.

References

Darren Lumbroso. (2017). 1-Risk KnowledgeIn:12 - Coastal Surges. Floods. 209–223.

Dawson, R. J., Dickson, M. E., Nicholls, R. J., Hall, J. W., Walkden, M. J. A., Stansby, P. K., Mokrech, M., Richards, J., Zhou, J., Milligan, J., Jordan, A., Pearson, S., Rees, J., Bates, P. D., Koukoulas, S. and Watkinson, S. R. (2009). Integrated analysis of risks of coastal flooding and cliff erosion under scenarios of long term change. *Climatic Change*. **95 (1–2)**: 249–288. doi:10.1007/s10584-008-9532-8.

Food and Agriculture Organization of the United Nations. (1998). Part B. Integration of Agriculture into coastal area management. Integrated Coastal Area Management and Agriculture, Forestry and Fisheries, *FAO guidelines*.FAO, pp. 256.

Michael, J. A. (2007). Episodic flooding and the cost of sea-level rise. *Ecological Economics*. **63**: 149–159. doi:10.1016/J.Ecolecon.2006.10.009.

Nicholls, R. J. (2002). Analysis of global impacts of sea-level rise: A case study of flooding. *Physics and Chemistry of the Earth, Parts A/B/C*. **27 (32–34)**: 1455–1466. Bibcode:(2002) PCE....27.1455N. doi:10.1016/S1474-7065(02)00090-6.

Nicholls, R. J., Wong, P. P., Burkett, V. R., Codignotto, J. O., Hay, J. E., McLean, R. F., Ragoonaden, S. and Woodroffe, C. D. (2007). Coastal systems and low-lying areas. In Parry, M. L., Canziani, O. F., Palutikof, J. P., Linden, P. J.

and Hanson, C. E. (eds.). *Climate Change (2007): Impacts, adaptation and vulnerability. Contribution of Working Group II to the Fourth Assessment Report of the Intergovernmental Panel on Climate Change*. Cambridge University Press. pp. 315–357.

Reynolds, M., Cairns, J., Stirling, C., Low, J., Campos, H. and Wassmann, R. (2017). Stress tolerant varieties to counter climate change. In: Dinesh D, Campbell B, Bonilla-Findji O, Richards M (eds). 10 best bet innovations for adaptation in agriculture: A supplement to the UNFCCC NAP Technical Guidelines. CCAFS Working Paper no. 215. Wageningen, The Netherlands: CGIAR Research Program on Climate Change, Agriculture and Food Security (CCAFS). Available online at: www.ccafs.cgiar.org.

Suarez, P., Anderson, W., Mahal, V. and Lakshmanan, T. R. (2005). Impacts of flooding and climate change on urban transportation: A system wide performance assessment of the Boston Metro Area. *Transportation Research Part D: Transport and Environment*. **10 (3)**: 231–244. doi:10.1016/j.trd.2005.04.007.

2

Farming System Approaches for Climate Resilience in Coastal Agriculture

R.M. Kathiresan
Department of Agronomy, Faculty of Agriculture
Annamalai University, Annamalainagar 608 002, Tamil Nadu

Introduction

Global warming directly reflects on rising sea levels due to melting of ice caps and natural expansion of sea water as it becomes warmer. Consequently, areas adjoining the coast and wetlands could be frequently flooded and the distribution pattern of monsoon rains may alter, through more intense downpours, storms and hurricanes. The meteorological data available at the Annamalai University, for the tail end of the Cauvery river delta region of Tamil Nadu State, India, shows that the average annual rainfall during the last 10 years segment has increased by 233 mm compared to the average of the previous 10 years segment (1588 mm and 1355 mm, respectively). In contrast, annual evaporation has reduced by 453 mm (2153 mm and 1700mm, respectively) (Kathiresan, 2011). The year 1990 was the hottest year in the last century with all other five of the warmest years in the century falling within the last 22 years. This trend indicates that flash floods in wetlands and droughts in uplands have become more frequent. Such a changing climate warrants reinforcement of climate resilience in farming to protect the livelihoods of farmers and the biodiversity. The sub-project of the National Agricultural Innovation Project implemented by Annamalai University in Tamilnadu, sponsored by ICAR has identified three types of clusters in each of the four disadvantaged districts, based on the prevailing agro-ecology *viz.* wetland clusters, rainfed upland clusters and coastal shore farming clusters. The constraints for farming and livelihoods in these clusters are identified as frequent floods, inundation, drought, weed problems, crop failure, and lack of diversification (Kathiresan, 2009a and Kathiresan, 2010). Farming systems that would be resilient to such climatic and farming constraints designed through rigorous institutional and on-farm experiments (Kathiresan, 2007; Kathiresan, 2009b and Kathiresan, 2012) were disseminated for adoption by the farmers in these clusters.

Materials and Methods

Four of the twenty nine districts of Tamil Nadu are identified by the Planning Commission of India as to have been disadvantaged considering the socio-economic, agro-ecological and bio-climatic indicators. Three of them *viz.*, Villupuram, Cuddalore and Nagapattinam lie on the coast. These districts are disadvantaged in terms of socio-economic conditions such as lower literacy (61 per cent as against the state average of 73.5 per cent), higher population of SC and ST population (28.09 per cent as against the states' figure of 20 per cent) and a vast majority of the farming community in these tracts are small and marginal farmers (72 per cent) with less than 2 hectare land holdings. In these districts wetlands characterised by monsoon dependant rice farming, with standing water and transplanted mode are predominant. However, interior parts that face scanty rainfall, offer scope for upland farming and coastal regions with fishing as main occupation are also prevalent. Accordingly three different agro-ecological clusters *viz.* wetland, upland and coastal clusters each with three villages were selected for the research. In total, nine clusters with 900 participating farmers were involved in the study, of whom 300 were involved in transfer of technology in each of the three types of clusters *viz.* wetland clusters, upland clusters and coastal clusters, during 2009-2012.

Wetland Clusters

Integrated Rice + Fish + Poultry farming system

Fish poly culture with Catla, Roghu, Mrigal, Common Carp and Grass Carp in equal proportions of a stocking density of 2000 finger lings ha^{-1} was taken up in trenches running along the border of rice fields on one side, with a dimension of 1 × 0.5 m, occupying 10 per cent rice area. Broiler birds @ 1 bird/10 m^2 of rice area (1000 birds ha^{-1}), were housed in cages that could accommodate a maximum of 20 birds (6′ × 4′ of floor space and a height of 3′), that were installed in the fields using four concrete posts of height 8′, 4′ buried inside the field and 4′ protruding above, lifting the cages above the crop canopy. The bottom of the cages were made of wire mesh (0.5 sq. inch) so as to leave the broiler waste, straight to the rice field wherein a 10 cm water column was maintained, allowing the poultry waste to get dissolved that served both as manure to the field as well as feed for the fishes. This excluded the need for collecting the poultry waste and applying it to the rice field, the task of which could be laborious and cumbersome, besides the scope for some wastage. The fishes reached out of the trenches into the standing water column in the rice fields during morning and evening hours, when the water temperature were tolerable and fed on the insect pests, weeds etc. (Fig. 1). Within a rice season in a farming year, three to five generations of poultry rearing with vencob breed were, taken up. The poultry birds imparted climate resilience during flash floods in the rice cropping reason that washed away the crop, by virtue of offering monetary returns, unaffected by floods.

Fig. 1. Individual Field View of Rice + Fish + Poultry Farming System

Upland Clusters

Integrated Goat + Crop Farming System

This technology involved rearing desi goats and using them for manuring as well as plant protection in crops that were grown during the succeeding cropping season. Under traditional goat rearing mode, farmers were rearing goats, that were feeding on exclusively herbs and vegetation available in the social and ranching sites. In this proposed project intervention, farmers were trained to rear the goats, allowing them to graze on the weed vegetation (mostly perennial grasses like *Cynodon dactylon* (L.) Pers. and sedges like *Cyperus rotundus* Linn.) that predominated the cropped lands during the off-season. Simultaneously, collecting the goat manure during the off-season and incorporating them for the crops (millet/vegetable/flower crop) during the rainfed seasons greatly complimented the crop by virtue of improved organic matter, soil nutritional status, pest, disease and weed control (through depletion of soil weed seed bank and suppression of weeds and alternate hosts of pests. However, these goats (reared @ 4/acre or 10/ha) were fed with tree toppings and other freely and easily available forages during the cropping season.

Coastal clusters

Integrated seaweed culture in coastal farming

This technology for seaweed culture involved growing of the seaweed *Gracilaria edulis* (S.G. Gmelin) P.C. Silva by raft or floats method. Fragments of this native sea weed were collected and used as seed. The floating devise was a simple frame made of bamboos durable in sea water. With bamboo, a 3m^2 frame was formed and 15 to 20 cross lines of rope were tied. In each cross line 20 seed fragments of sea weed were tied. The raft was made to float at 50 cm below sea water surface, with the help of floats made of foam or plastic buoys or bags filled with coconut husks. The frame was also anchored to bottom using bamboo poles. From 1 ha, 100 tons of fresh sea weed bio-mass, resulting in 10 tons of dry sea weed or 2 tons of agar annually were obtained from each harvest cycle within 60 days.

Results and Discussion

Wetland Clusters

The baseline survey of the project indicated that the gross household annual income in Wetland clusters is ₹31,822.11. Increases in income for these three districts through the adoption of Rice + Fish + Poultry farming are presented in Table 1. The increase in Gross household income in Villupuram district was

Table 1. Livelihood enhancement in wetland clusters

Particulars	Villupuram	Cuddalore	Nagapattinam	Weighted mean of the three districts
No. of poultry bird rearing	7	5	5	5
Average meat yield/bird (kg)	2.40	2.50	2.10	2.3
Average meat yield/ household (kg)	336	250	210	2.65
Cost of meat ₹/kg	100	110	90	100
Gross return from poultry (₹)	33,600	27,500	18,900	26,666
Cost of production of poultry bird (₹)	9,900	5,700	7,100	7,566
No. of Fish rearing	2	1	1	1
Fish yield/household (kg)	120	75	75	90
Fish cost ₹/kg	70	90	80	80
Gross return from fish (₹)	8,400	6,750	6,000	7,050
Cost of production of fish (₹)	900	500	500	633
Total net return, household/ year (₹)	31,200	28,050	17,300	25,516
Livelihood enhancement (%)	98	88	54	80

Table 2. Manurial Addition from Poultry Voidings

Districts	Age of birds (in days)	Voiding by a bird/day (g)	Total quantity of voiding added for 5 cents of rice/year (Kg)	Nutrients added (Kg)			Recommended quantity of FYM for 5 cents rice (Kg)	Nutrients added (Kg)		
			1.73% N, 0.85% P_2O_5 & 0.38% K_2O	N	P_2O_5	K_2O		N	P_2O_5	K_2O
Villupuram	15	38.0	392.00	6.78	3.33	1.48	600.00	3.00	1.50	3.00
	30	71.5								
	45	77.3								
Cuddalore	15	39.2	228.00	3.94	1.93	0.86	200.00	1.00	0.50	1.00
	30	71.8								
	45	78.5								
Nagapattinam	15	37.8	279.45	4.83	2.37	1.06	400.00	2.00	1.00	2.00
	30	70.6								
	45	77.9								
Weighted mean of the three districts	-	-	300	5.18	2.54	1.13	400.00	2.00	1.00	2.00

₹31,200 which was accounting for 98 per cent, the highest, as the number of broiler rearing spread over three crops of rice was seven. The increase in gross household income in Cuddalore district was ₹28,050, that contributed only 88 per cent increase. This was because of the fact that water availability in the wetland cluster of this district was not permitting more than one crop and hence, only four broiler rearings alone were possible. However, the farmers were enthusiastic about the intervention as is evident from one broiler rearing that extended outside the three that are possible during the single crop of rice. The increase in gross household income was the least of ₹17,300 that made only 54 per cent increase with in an year in Nagapattinam district. Though two crops of rice were grown and poultry rearings for five generations were taken up, the meat yield and market prices were comparatively lower than that experienced at Cuddalore. The Manurial output from broiler birds in rice are furnished in Table 2. The results indicate that addition of poultry manure in five cents of rice area added nutrients more than the quantity that could have been possible through the normally recommended dose of Farm Yard Manure. Higher nutrient addition through poultry manure compared to other organic sources in rice was already observed in institutional and on-farm experiments. Pest incidence in rice as shown in Table 3, was also reduced due to integration of the fish culture and poultry components, because of the feeding habits of fishes that suppresses the egg masses, larvae and alternate weed hosts of pests. Pest suppression and livelihood enhancements were experiencing incremental enhancements from those observed in previous year results.

Table 3. Rice + Fish + Poultry and Pest Incidence in Rice

Districts	Leaf Damage in % on 40 DAT		*Nilaparvata lugens* Population on 7 DAT	
	Rice Alone	Rice + Fish + Poultry	Rice Alone	Rice + Fish + Poultry
Villupuram	21.0	17.0	14.0	10.0
Cuddalore	23.0	18.0	11.0	8.0
Nagapattinam	17.0	14.0	15.0	11.0
Weighted mean of the three districts	20.3	16.3	13.3	9.6

The striking success of this Rice + Fish + Poultry farming system made 392 other farmers (other than the 838 identified development partners) to adopt this in their holding. Further 12 of the identified development partners extended the technology from the project supported 200 m^2 area to half an acre (2000 m^2) of their holdings. Further, the State Planning Commission of Tamilnadu, on requests from farmers who were consulted, at their regional consultative meetings, invited an exclusive presentation of the project interventions at the State Planning Commission Chennai, where in all the secretaries and executives of line departments participated.

Upland clusters

In the upland clusters, integrating goats rearing with crops of farmer's choice like vegetables or flower crops or millets, in sequence, enhanced the livelihood of small and marginal farmers. The baseline survey indicated that annual income of farmers in upland clusters was ₹30,398, in all the three districts. The average livelihood enhancement was the highest in Villupuram district (Table 4), as the reproduction from two goats given to the farmers, at the commencement of the project, resulted in eleven goats in total after 36 months with nine of them becoming saleable. In other two districts, though the reproduction rate on an average is comparable, the weight increase in kids is slow because of lesser availability of the loppable fodder during cropping season and weed vegetation available for grazing during the off-season. This is reflected on the fact that farmers of these districts *viz.*, Cuddalore and Nagapattinam had only seven additional goats that were salable by 36 months duration that contributed for comparatively lower livelihood enhancement. However, the reduction in weed biomass in the farmers fields because of grazing by goats in the off-season. (Fig. 2) was higher in Cuddalore and Nagapattinam districts compared to Villupuram that could be attributed to closer grazing of goats for want of excessive or adequate flushes of weed vegetation in the off-season in these two districts compared to Villupuram.

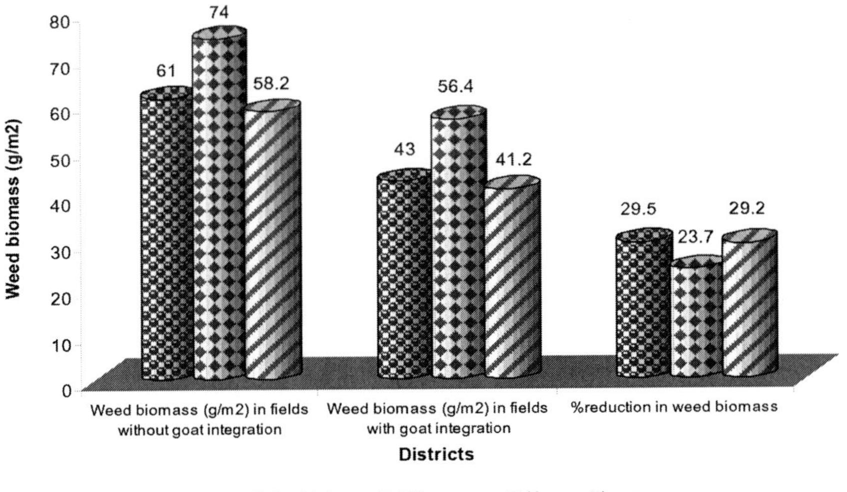

Fig. 2. Weed suppression due to goat grazing in upland clusters

In the upland clusters, the goats given to 300 farmers (two for each development partner) multiplied to 6000 at the end of four years (March 2012) after the sale of ten goats by a development partner (on an average). Further, these 6000 goats multiplied to 10,000 goats at the end of March 2014. The manurial addition from these goats and weed control by virtue of grazing on weed vegetation during the off season greatly complimented crop production. The details are given in Table 4.

Table 4. Livelihood enhancement in upland clusters

Districts	Number of goats sold on reproduction	Return from sale of goats (₹)	Maize yield in Kg/ha	Enhancement in livelihood (%)
Villupuram	Ten	25,000.00	2175	82
Cuddalore	Ten	23,000.00	1850	75
Nagapattinam	Ten	20,000.00	1920	65
Weighted mean of the three districts	Ten	22,666	1981	74

Coastal Clusters

In shore farming clusters, culture of seaweed *Gracilaria edulis* improved the annual household income of the participating development partners (Table 5). The baseline survey value indicated that their annual income was ₹20,526. Villupuram district experienced the highest of 68 per cent of livelihood enhancement, as the number of harvests was five/year, made possible by lesser dilution of the brackishness of backwater during the monsoon seasons. Cuddalore was next in order (with even four harvests/year) as the yield of seaweed biomass is 220 kg/raft which was slightly higher than the other two districts where the yields were only 200 kg of fresh seaweed/raft. The enthusiasm of development partners was lower in Nagapattinam district that accounted for only three rafts compared to four rafts used for seaweed culture in other districts. This solely contributed for the least enhancement of livelihood (39 per cent).

Table 5. Livelihood enhancement in coastal clusters

Location	No. of rafts	No. of harvests	Total yield of dry seaweed (kg)	Return/ household (₹)	Livelihood enhancement (%)
Villupuram	4	5	700	14,000.00	68
Cuddalore	4	4	670	13,400.00	65
Nagapattinam	3	4	400	8,000.00	39
Weighted mean of the three districts	3	4	590	11,800.00	57

Conclusion

Integrating other farm enterprises with the traditional cropping pattern in small holder farms offers good scope for climate resilience, enhancing livelihoods and in imparting environmental security. This is evident from enhanced livelihoods by 80% in wetland clusters, 74% in upland clusters and 57% in coastal clusters. Further, reduced pest incidence in rice and weed incidence in upland crops in the absence of inorganic plant protection and addition of poultry manure to a tune of 1026.48 kg/ha per year on an average from all three wetland clusters also support this observation.

Acknowledgement

The funding support from ICAR-NAIP Component-III, Sustainable Rural Livelihood Security for taking up this research is gratefully acknowledged. The co-operation rendered by consortium partners *viz.* Dhan Foundation, Vedapuri KVK and BMT-KVK of Tamilnadu in transfer of technology is appreciated with gratitude.

References

Kathiresan, RM. (2007). Integration of elements of farming system for sustainable weed and pest management in the tropics, *Crop Protection*, **26:** 424-429.

Kathiresan, RM. (2009a). Sustainability through Integrated Farming Systems in small holder farms of Tamilnadu State of India. *Proceedings of Farming System Design Symposium,* pp. 217-218. 23-26 August (2009). Monterey California, USA.

Kathiresan, RM. (2009b). Integrated farm management for linking environment, *Indian Journal of Agronomy*, **54(1):** 9-14.

Kathiresan, RM. (2010). Spatial and Temporal Integration of Component Enterprises in Small Holder Farms of India for Sustainability in Farming and Rural Livelihoods. *Proceedings of Ninth European IFSA Symposium.* pp. 2123-2128. 4-7 July (2010). Vienna, Austria.

Kathiresan, RM. (2011). Utility TAG, Farming elements and ITK for sustainable management of weeds in changing climate. *Proceedings of 23rd Asian-Pacific Weed Science Society Conference*, pp. 228-238. 26-29 September (2011). Cairns, Australia.

Kathiresan, RM. (2012). Climate resilient integrated farming systems for sustainability and livelihood enhancement. *Proceedings of 3rd International Agronomy Congress,* pp. 955-957. 26-30. November (2012). New Delhi.

3

Molecular Breeding for Salt and Submergence Tolerance in Rice

S. Thirumeni, K. Paramasivam and J. Karthick
Pandit Jawaharlal Nehru College of Agriculture and Research Institute
Karaikal 609 603, Puducherry (*U.T*).

Introduction

Food supply is a major concern in the near future as the human population touches 9 billion in 2050 and therefore, more food has to be produced. This has to be achieved in the face of climate change amounting to constant pressure of biotic and abiotic stresses. However, Swaminathan (2007) opined that science and technology can play a very important role in stimulating and sustaining an evergreen revolution leading to long-term increases in productivity without associated ecological harm. Rice is considered as global grain since it is consumed by more than half of the world population. In India rice supplies calorie requirement to more than 70 per cent of its population. This is evident from the fact that rice is cultivated in all the 30 states and 5 union territories of the Indian subcontinent of diverse ecologies ranging from below sea level to mountainous regions. Though India tops in rice acreage (43.61 mha) its productivity of 3.19 t/ha (rough rice) is far below world average of 4.12 t/ha and that of Asia's average of 4.19 t/ha (Shobharani *et al.*, 2010). This low productivity is mainly due to unfavourable ecologies of the varied ecosystems which, rice encounters during critical stages of its growth. Salinity and submergence are major abiotic stresses, next to drought, limiting rice production in South East Asia including India.

In India nearly 8 mha are salt affected of which 2 mha are coastal saline and 3.4 mha are sodic (Singh, 1994). Rice is predominant crop in these areas especially in vast stretches of eastern and western coasts of India. The presence of salts, in excess to plant growth, in the soil solution makes the soil detrimental to crop growth. The source of the excess salt may be from soil or irrigation water. The content and composition of salts (ions) in tandem with types of soil and weather conditions prevailing during the crop period decides the severity of salt stress. Similarly, flash floods or short-term submergence regularly affect rice growing areas in South and Southeast Asia. Even more favorable irrigated areas experience flooding problems during the monsoon season. It is also prevalent in rainfed lowlands where most of the present day rice varieties do not survive complete submergence for more than a week. In India 30% of the rice growing area is prone to such menace resulting in severe yield losses (Neeraja *et al.*, 2007).

Green revolution stand testimony to the power of plant breeding, which will continue to play crucial role in agriculture as increase in food production is realizable mainly through development of improved crop varieties and hybrids coupled with better management practices. Since then, expectations from plant breeding, for new breeds of crops that sustains higher/stable yield in stress prone areas, run high among public and policy makers. To meet these challenges breeders brought in novel genes/alleles for abiotic stress tolerances into high yielding semi-dwarf background prevalent in farmers' fields now. Tolerance of rice land races to a variety of abiotic stresses highlights the success of early farmers (plant selectors). However, it is due to recent advances in genetics and marker techniques that plant breeders have begun to develop new methods for developing stress tolerant and high yielding varieties (Thomson *et al.*, 2010).

In this chapter, the application of molecular markers as tools for dissecting salt and submergence tolerance traits in rice, through QTL mapping, and its subsequent use in MAS especially MABC are discussed. Also case studies pertaining to improving salt and submergence tolerances in rice are presented so as to showcase the potential of QTL mapping and MAS.

Molecular breeding for salt and submergence tolerance.

Though we could increase food production through green revolution with little or no knowledge of the genes underlying the genetic variability exploited by breeders for crop improvement, this is, however, insufficient now to produce more food from fragile environments like salt stressed soils and submergence prone areas (Collins *et al.*, 2008).Constraints in breeding for stress tolerant rice include poor understanding of the mechanisms underlying tolerance,complexity of the trait,lack of efficient selection criteria, absence of reliable& repeatable screening methodology and variation of tolerance with ontogeny (Gregorio *et al.*, 2002).

Crop tolerance to stresses is reflected by an interrelated set of complex physiological and morphological traits, each with an intricate regulatory system (Thomson *et al.,* 2010). By integrating physiological and genetic changes, the underlying molecular mechanisms can be deciphered. This opens up way for a more targeted breeding approach to develop stress tolerant rice varieties. This breakthrough is made possible through the use of molecular marker technology. Using these highly polymorphic markers saturated linkage maps have been prepared thus enabling identification of chromosomal regions associated with many abiotic stresses including salt and submergence tolerances through Quantitative Trait Loci (QTL) analysis which, hitherto, was not possible through classical genetic approach. The marker (s) linked to the trait of interest, identified through QTL analysis, is (are) to be used as molecular tag to identify plants (selection) in breeding populations, popularly referred as Marker Assisted Selection (MAS). This method of selection, based directly on gene (s) of interest/QTL, is also referred to as molecular breeding. To practice of molecular breeding, mapping of the target traits, through QTL analysis, QTL confirmation and marker validations are pre-requisite.

QTL Mapping

Most of the economically important traits of crop plants are quantitatively inherited and, therefore, analysed using classical quantitative genetic models assuming(1) unknown number of genes are determining a quantitative trait, (2) each gene is having small effect, (3) gene effects at loci are additive, (4) genes are showing G X E interaction. Further, estimates of genetical component of variation were obtained assuming equal effects, complete dominance, no epistasis and no linkage. Proponents of biometrical analysis of quantitative traits assumed this because at that time there were no way in which one could determine the location and effect of each gene determining quantitative variation. Now with the advent of the molecular markers, the scenario has changed and one can now precisely and accurately locate and measure the effect of the individual loci. Such a chromosomal region linked or associated with a marker gene which affects a quantitative trait is defined as quantitative trait locus (QTL). Not only the number of QTLs determining a quantitative trait can be counted but also estimate the gene effects without any assumptions mentioned above. Thus, QTL mapping throw light on genetics of complex traits like yield, quality, tolerance to drought, cold, heat, salt, submergence etc.

QTL analysis comprises (Collard *et al.*, 2005); (Figure 1 & 2) (1) Development of mapping population, (2) Generation of marker data using the mapping population, (3) Construction of linkage map using the marker data, (4) Precise phenotyping for the trait under investigation and (5) Identifying the chromosomal region correlated to the phenotype using molecular markers (employing various mapping methods/softwares).

For abiotic stress tolerance, it become possible to test different physiological components and compare the QTL locations for these stress tolerances with

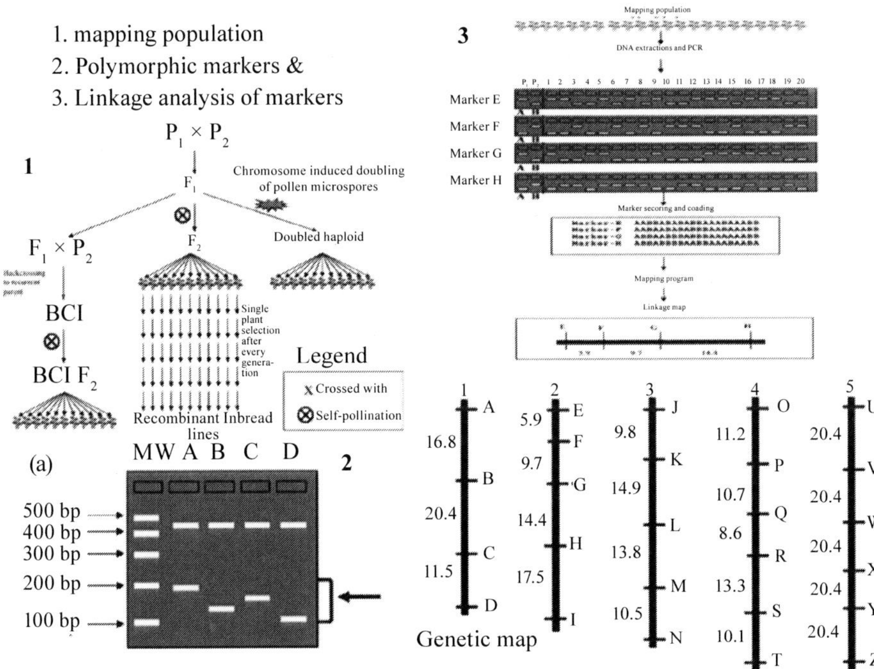

Figure 1. Steps in genetic map construction (adopted from Collard *et al.*, 2005)

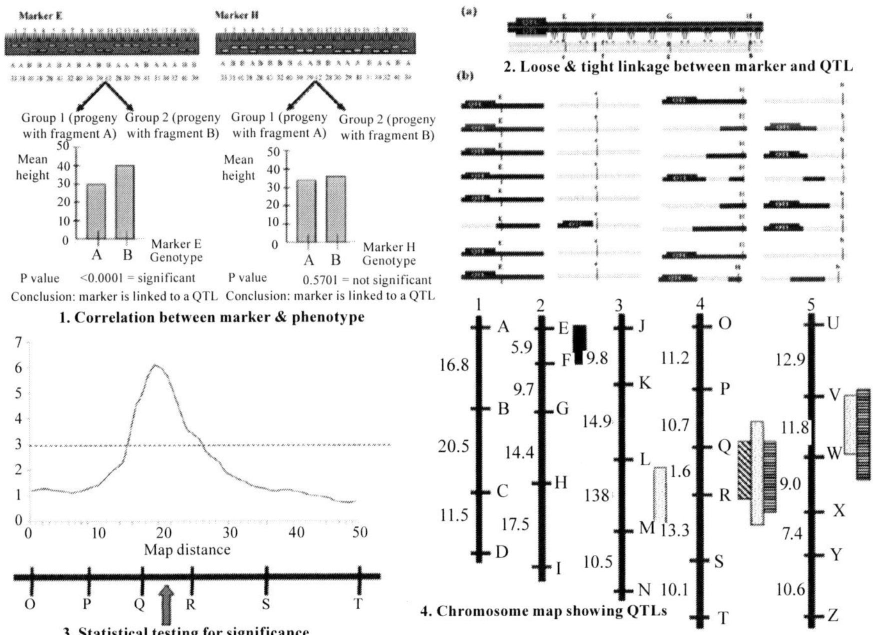

Figure 2. Steps in QTL Mapping (adopted from Collard *et al.*, 2005)

the QTLs for tolerance or yield under stress to identify the causal factors (Thomson *et al.*, 2010). In addition, use of immortal mapping populations such as Recombinant Inbred Lines (RILs) or Introgression Lines/Chromosome segment substitution lines (ILs/CSSLs) enables the conduct of phenotyping in replicated trials over locations. This ultimately identifies QTLs which are specific to different stress levels. Once large effect QTLs for stress tolerance are identified, these can be characterized using ILs, paving way for further cloning of the QTL to identify the causal gene.

Having complete sequencing data in rice on hand paved way for preparing a universal consensus map that can bring together genetic mapping data from various sources into a single physical map based on the DNA sequence. Apart from that, it provides many new markers across the genome which is essential for fine mapping. These advances have made cloning of QTL to isolate the causal gene easier (Thomson *et al.*, 2010). Some of the major QTLs for abiotic stresses fine mapped and cloned earlier were that of *Saltol* QTL (SKC) for salt tolerance (Ren *et al.*, 2005) and *Sub1* QTL for submergence tolerance (Xu *et al.*, 2006).

Marker-Assisted Selection (MAS)

The ultimate aim of QTL mapping is to use markers tightly linked to major QTLs for abiotic stress tolerance for indirect selection. DNA markers are reliable selection tools because they are stable, not influenced by environment and easy to score in a laboratory. When compared to traditional phenotypic selection, marker based selection (indirect selection) offers great advantages in variety development which includes increased reliability of the trait under selection, improved efficiency in selection and reduced cost especially with certain traits whose phenotyping is very cumbersome and costly (Collard *et al.*, 2008). Of the many types of MAS, marker assisted back crossing (MABC) has been widely used in many crop improvement programmes so far. The basis of a marker-assisted backcrossing (MAB)is to transfer a specific allele at the target locus from a donor line to a recipient line while selecting against donor introgressions across the rest of the genome. Through the use of co-dominant markers, the speed of the selection process can be increased, thus enhancing genetic gain per unit time (Hospital, 2009). The merits of MAB are: (1) efficient foreground selection for the target locus, (2) reduction of linkage drag surrounding the gene/QTL through recombinant selection, (3) efficient background selection for the recurrent parent genome, and (4) rapid selection of new genotypes with desired traits. The effectiveness of MAB depends on the availability of closely linked markers and/or flanking markers for the target locus, the size of the population, the number of backcrosses and the position and number of markers for background selection (Hospital, 2009).

As of now, MABC has been successful in transferring traits whose expression is controlled by a single gene or by a gene that controls maximum phenotypic variation of the trait. An example of a single QTL introgression into several popular rice varieties is explained in Figure 3.

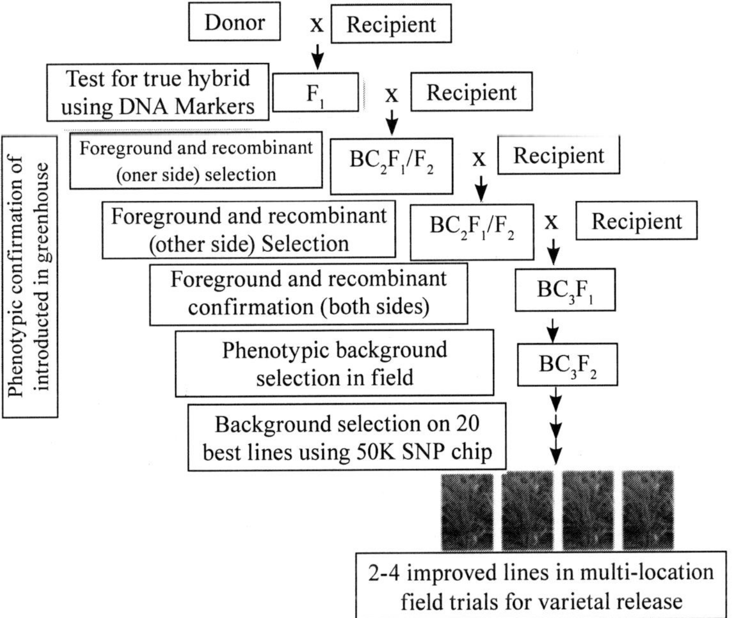

Figure 3. Marker-assisted backcross breeding procedure for the transfer of major QTLs for abiotic stress tolerance into popular rice varieties (Singh *et al.*, 2016).

Case studies

Marker Assisted Backcross Breeding (MABC) for salt tolerance

Classical breeding has supported the development of improved salt-tolerant genotypes up to a certain degree wherein selection for salt tolerance was focussed on yield and few of its critical determinants like tiller number, grain number and grain weight under salt stress and normal environment. With the advancement in the knowledge of physiological mechanisms related to salt tolerance key traits identified are salt exclusion, low uptake, compartmentation of toxic ions in structural and older leaves, tissue tolerance Reactive Oxygen Species (ROS) scavenging and water use efficiency. To go further in improving salt tolerance of rice varieties, Flowers (2005) emphasized the traits to be pooled include Na+ exclusion, K+/Na+ discrimination, ion retention in sheaths, tissue tolerance, ion partitioning in different leaves, osmotic adjustment, enhanced vigor, water use efficiency and early flowering. Since markers have been found as linked to some of these traits and already cloned genes, MAS offers the possibility to pyramid efficiently the QTLs or genes contributing to the different plant tolerance strategies.

The detection of QTLs (Quantitative Trait Loci) has enabled larger progress in understanding the genetic control of the trait and the underlying physiology. Over past two decades various researchers mapped QTLs for salt tolerance and its components in different mapping populations which show majority of these QTLs are robust, being identified from multiple donors (Platten *et al.*, 2013). Thus, the major QTL *SKC/Saltol* is reported in almost all studies involving Nona Bokra, Pokkali, Kala rata, Patnai and cheruvirrippu as donors. In the same way most of the QTLs are shared between several donors (Table 1). This QTL explains around 40% of phenotypic variation for seedling stage sat tolerance. This reflects the shared origins of many donors in India and Bangladesh and their common physiological mechanisms (dominated by Na exclusion). Further QTL studies on physiological traits, other than survival and scores and Na/K contents and ratio, are scanty. Thus, there is scope for identification of additional QTLs for ROS scavenging, tissue tolerance etc., in the germplasm especially outside India. A segment, around 10–12 Mb region, of the short arm of chromosome 1 concentrates a large number of QTLs (Figure 4) indicating many studies confirmed the location of *Saltol* QTL (Negrao *et al.*, 2011).

Table 1. QTLs common to multiple mapping populations (Platten *et al.*, 2013)

Chromosome	Position (Mb)	Tolerant donors
1	11	Pokkali, Nona Bokra, ChikiramPatnai, Kala Rata, FL478, Cheriviruppu
1	39	Nona Bokra, FL478, Capsule, Cheriviruppu, Chikiram, Patnai, Gihobyeo, Nipponbare
2	28	Capsule, Kala Rata, FL478, IR64, Kasalath
3	5	Pokkali, IR64, FL478, ChikiramPatnai, Capsule
3	30	Pokkali, Kasalath, Kala Rata, FL478, Cheriviruppu, Capsule
4	23	Nona Bokra, Boilam, ChikiramPatnai, IR64, JX17
5	24	Capsule, Kala Rata, IR64
6	5	FL478, IR64, JX17
6	22	Nona Bokra, Pokkali, FL478, JX17, Tarommahalli
7	4	Nona Bokra, Nipponbare, IR64, Boilam
9	15	Nona Bokra, Pokkali, IR64
12	8	Pokkali, Kala Rata, FL478, Capsule
12	25	Pokkali, JX17, FL478

Identification of many QTLs provide more opportunities for breeding programmes to improve salt tolerance in mega rice varieties through MABC.

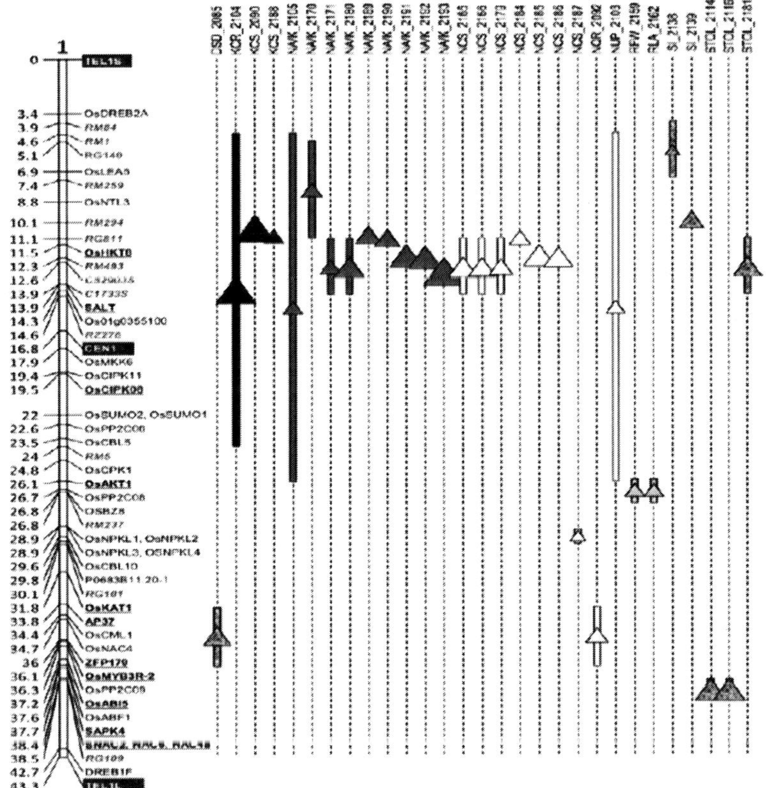

Figure 4. Map showing Saltol QTL on chromosome 1 reported in many studies (Negaraoet 2011).

Current efforts in various institutions are directed towards marker-assisted back crossing programs to introgress the favorable alleles for tolerance QTLs into elite rice lines for improving seedling stage tolerance. At IRRI, Philippines, mega varieties like IR 64, BR 11, BRRI Dhan 28, BRRI Dhan 29 (Thomson *et al*., 2010) and Rassi (Bimpong *et al*., 2016) were improved by introgression of Saltol QTL. In India at IARI, New Delhi, Pusa basmati 1121 was improved for seedling stage tolerance in the same way. In another major initiative Indian government, through Department of Biotechnology, funded a project captioned "From QTL to Variety: Marker Assisted Breeding of Abiotic Stress Tolerant Rice Varieties with Major QTLs for drought, submergence and salt tolerance" to develop stress tolerant rice varieties. In this project popular rice varieties namely ADT 45, CR 1009, Gayathri, MTU 1010, PR 114, Pusa 44 and Sarjoo 52 are introgressed with Saltol QTL for improving salt tolerance (Singh *et al*., 2016). Of these ADT45 Saltol and MTU1010 Saltol Introgression lines are nominated for multi location testing.

Though many MABC programmes were attempted by various laboratories on different mega rice varieties, none of them resulted in release of salt tolerant variety. The reason being required level of tolerance is not achieved through only one major QTL and therefore, many more robust QTLs are to be pyramided in the same genetic background to enhance the level of tolerance. Also QTLs at seedling and reproductive stages are to be combined as tolerance during these stages are weakly associated.

Breeding for Submergence tolerance

Water though essential for growth, excess during monsoon results in submergence or water logging which is very damaging and incurs yield loss. Submergence predispose plants to the stresses of low light, limited gas diffusion, effusion of soil nutrients, mechanical damage, and increased susceptibility to pests and diseases. Rice is a semi-aquatic species that is typically cultivated under partially flooded conditions. However, flash flooding can cover the entire plant for prolonged periods, and most of rice varieties die within seven days of complete submergence (Xu *et al.*, 2006 and Bailey- Serres *et al.*, 2010) and only a limited number of rice varieties survives under submergence. Submergence can damage rice fields at any stage of crop growth. Germination is greatly affected indirect-seeded rice after heavy rainfall while during vegetative growth, damage can cause total loss if submergence is longer than a week. The extent of submergence damage depends on environmental conditions: higher temperatures, higher turbidity, and lower solar radiation etc.

Two varieties: FR13A and FR43B, reported to be "flood resistant," (Richharia and Govindaswami, 1966) were pure-line selections from traditional landraces (Dhalputtia and Bhetnasia in the case of FR 13A & FR43B respectively). Since then, it has been the most widely used donor of tolerance in breeding programmes.

While Mishra *et al.* (1996) reported genetics of submergence tolerance to be controlled by a single dominant gene, through QTL mapping technology a major QTL (*Sub1*) explaining about 70% of phenotypic variation (for the trait) has been identified and fine mapped on chromosome 9 in FR13A (Xu and Mackill, 1996) and many other studies confirmed this major QTL(Bailey-Serres *et al.*, 2010; Figure 5). Further additional QTLs from "FR13A" on chromosomes 1, 8, and 10 were also subsequently identified (Gonzanga *et al.*, 2016).

Three related ethylene response factor (ERF)-like genes at this Sub1 locus were identified, *Sub1A*, *B* and *C*, although *japonica* varieties and some *indicas* do nothave the *Sub1A* gene (Xu *et al.*, 2006). *Sub1A* and *Sub1C* were

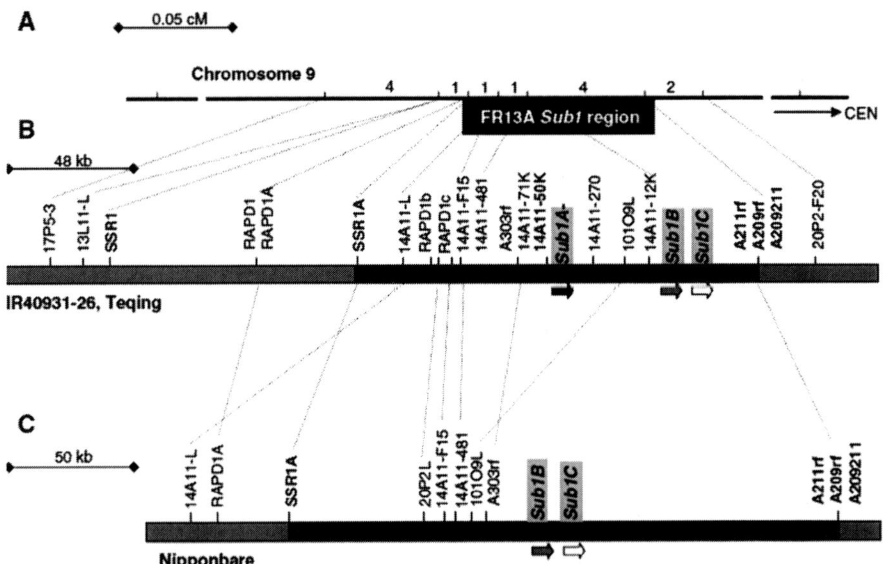

Figure 5. Genetic and physical map of the SUB1 locus on the long arm of chromosome 9 (Bailey-Serres *et al.*, 2010)

up-regulated by submergence and ethylene (Fukao*et al.*, 2006). *Sub1A* was strongly induced in the tolerant varieties in response to submergence, whereas intolerant varieties had weak or no induction of the gene. Over expression of *Sub1A* conferred submergence tolerance in an intolerant *japonica* variety and down-regulation of *Sub1C* (Xu *et al.*, 2006).

At IRRI, a programme to introgress SUB1 QTL into Six mega-varieties namely Swarna, Samba Mahsuri, BR11, IR64, CR1009 and TDK1, using marker-assisted back crossing (MABC) was initiated (Ismail *et al.*, 2013). Using this approach, a small genomic region containing the SUB1 locus was introgressed into each of the six modern high-yielding varieties within 2–3 years (Neeraja *et al.*, 2007; Septiningsih *et al.*, 2009; Iftekharuddaula *et al.*, 2011). More recently, introgression of SUB1 into two more varieties, PSB Rc18 from the Philippines and Ciherang from Indonesia, was also completed (Table 2). This illustrates that the availability of numerous Sub1 varieties with diverse genetic backgrounds will allow selection of Sub1 donors that are the most similar to any new variety to be introgressed with SUB1, and this shortened the breeding cycle considerably and reduced the cost of development.

Similar to Introgression of *Saltol* QTL into Indian rice mega varieties, through DBT funded "From QTL to Variety" project (Singh *et al.*, 2016), as much as ten popular rice varieties namely ADT 39, ADT 46, Bahadur, HUR 105, MTU1075, Pooja, Pratikshya, Rajendra mashuri, Ranjit and Sarjoo are introgressed with SUB1 QTL for improving submergence tolerance and are at different at levels of multilocation testing.

Table 2. Some popular SUB1 varieties developed through MABC

Recurrent parent	Country	Donor parent	Generation	Country and Year released
Swarna	India	IR 49830	BC3F2	India (2009), Indonesia (2010),Nepal, Bangladesh and Myanmar (2011)
IR 64	Philippines	IR 40931	BC2F2	Philippines & Indonesia (2009)
Samba mashuri	India	IR 49830	BC2F2	Nepal (2011)& India (2013)
TDK 1	Laos	IR 40931	BC3F2	-
BR 11	Bangladesh	IR 40931	BC2F2	Bangladesh (2010)
CR 1009	India	IR 40931	BC2F3	India (2016)
Ciherang	Indonesia	IR 64 Sub1	BC1F2	Indonesia (2012), Bangladesh (2013)
PSB Rc 18	Philippines	IR 64 Sub1	BC1F2	-

Conclusion

Crop improvement through breeding brings immense value relative to investment and offers an effective approach to improving food security. The importance of rice as staple food and its high sensitivity to salt and submergence warrant in depth study of tolerance mechanisms. The tolerance mechanisms and strategies adopted by those varieties seem to be quite diverse and polygenically controlled. Comprehensive maps integrating tolerance QTLs and candidate genes reinforce the relevance for breeding for stress-tolerance of certain chromosome regions where both co-localize. Overall, the better knowledge of the function and regulation of the responsive genes and their association to QTL regions will allow a more structured approach to breeding for stress tolerance. It was believed that molecular breeding could remove frontier between Mendelian and Statistical genetics (Crow, 1993). But this turned out to be not completely true (Hospital, 2009). Some traits, complex and quantitative, were dissected to simple mendelian traits (since few QTLs with large effect were identified as in the case of submergence tolerance) but many could not be as in the case of salt tolerance. However, modern genomics and biotech tools has the potential for second green revolution that would come about by combining the invaluable scientific methodology and products with conventional breeding techniques. Therefore, it remains to be seen how best genetics of stress tolerance is understood that eventually leads tobreed superior stress tolerant varieties.

References

Bailey-Serres, J., Fukao, T., Ronald, P.C., Ismail, A.M., Heuer, S., Mackill, D. (2010). Submergence tolerant rice: SUB1's journey from landrace to modern cultivar. *Rice* **3**: 138–147.

Bimpong,I.K., Baboucar, M., Mamadou, S. *et al.* (2016). Improving salt tolerance of lowland rice cultivar 'Rassi' throughmarker-aided backcross breeding in West Africa. *Plant science* **242**: 288-299.

Catling, D. (1992). Rice in Deep Water. International Rice Research Institute, Los Ba˜nos, Philippines, 542 pp.

Collard, B.C.Y., Jahufer,M.Z.Z., Brouwer,J.B.,and Pang.E.C.R.(2005). An introduction to markers, quantitative trait loci (QTL) mapping, and marker-assisted selection for crop improvement: The basic concepts. *Euphytica* **142**: 169–196.

Collard, B.C.Y., Mackill, D.J. (2008). Marker-assisted selection: an approach for pre-cision plant breeding in the 21st century. *Philos. Trans. R. Soc. B: Rev.* **363**: 557–572.

Collins,N.C., Tardieu, F. and Tuberosa, R.(2008). QTL and crop performance under Abiotic stress: Where do we stand? *Plant physiology*, **147**: 469-486.

Crow,J.F. (1993). Galton, Francis-count and measure and count. **Genetics,** 135 (1): 1-4.

Flowers, T.J. (2004). Improving crop salt tolerance. *Journal of Experimental Botany*. **55**: 113.

Fukao, T., Xu, K., Ronald, P.C., Bailey-Serres, J. (2006). A variable cluster of ethylene response factor-like genes regulates metabolic and developmental acclimation responses to submergence in rice. *Plant Cell.***18**: 2021–2034.

Gonzaga,Z.J.C., Carandang,J., Sanchez,DL, Mackill,DJ. and Septiningsih,EM. (2016). Mapping additional QTLs from FR13A to increase submergence tolerance in rice beyond SUB1.*Euphytica*. DOI 10.1007/s10681-016-1636.

Gregorio, G.B., Sanadhira,D., Mendoza,R.D., Manigbas,N.L., Roxas, J.P., and Guerta, C.Q.(2002). Progress in breeding for salinity tolerance and associated abiotic stress in rice. *Field Crops Research*,**76**: 91-101.

Hospital,F. (2009). Challenges for effective marker-assisted selection in plants. *Genetica*.**136**:303–310.

Iftekharuddaula, K., Newaz, M., Salam, M., Ahmed, H., Mahbub, M., Septin-ingsih, E., Collard, B., Sanchez, D., Pamplona, A., Mackill, D. (2011). Rapid and high-precision marker assisted backcrossing to introgress the SUB1 QTL into BR11, the rainfed lowland rice mega variety of Bangladesh. *Euphytica*. **178**: 83–97.

Ismail, A. M., Ella, E. S., Vergara, G. V., and Mackill, D. J. (2013). The contribution of submergence-tolerant (Sub1) rice varieties to food security in flood-prone rainfed lowland areas in Asia.*Field Crop Research*. http://dx.doi.org/10.1016/j.fcr.2013.01.007.

Mackill, D.J., Ismail, A.M., Singh, U.S., Labios, R.V., Paris, T.R.(2012). Development and rapid adoption of submergence-tolerant (Sub1) rice varieties. *Adv. Agron.* **115**: 03–356.

Mishra,S.B., Senadhira, D., and Manigbas, N.N. (1996). Genetics of submergence toleranceinrice. *Field Crop Research,* **46**: 177-181.

Moradi F., Ismail A.M., Gregorio G.B., and Egdane, J.A. (2003). Salinity tolerance of rice during reproductivedevelopment and association with tolerance at the seedling stage. *Indian Journal of Plant Physiology*, **8**: 105–116.

Neeraja, C.N., Maghirang-Rodriguez, R., Pamplona, A., Heuer, S., Collard, B.C.Y., Septiningsih, E.M., Vergara, G., Sanchez, D.L.M., Xu, K., Ismail, A.M., Mackill, D.J. (2007). A marker-assisted backcross approach for developing submergencetolerant rice cultivars. *Theor. Appl. Genet.* **115**: 767–776.

Negrão, S., Courtois,B., Ahmadi,N., Abreu,I., Saibo, N., and Oliveira, M. (2011). Recent Updates on Salinity Stress in Rice: From Physiological to Molecular Responses, *Critical Reviews in Plant Sciences*, **30(4)**: 329-377.

Platten, J., Egdane, J., Ismail, A. (2013). Salinity tolerance, Na+ exclusion and allele mining of *HKT1;5*in *Oryza sativa* and *O. glaberrima*: many sources, many genes, one mechanism? *BMC Plant Biol,* **13**: 32.https://doi.org/10.1186/1471-2229-13-32.

Richharia, R. H., and Govindaswami, S. (1966). *Rices of India*. Scientific Book Company, Patna, India.

Ren, Z. H., Gao, J. P., Li, L. G., Cai, X. L., Huang, W., Chao, D. Y., Zhu,M. Z,, Wang, Z. Y., Luan, S., and Lin, H. X. (2005). A rice quantitative traitlocus for salt tolerance encodes a sodium transporter. *Nat. Genetics,* **37**:1141–1146.

Septiningsih, E.M, Ignacio, J.C.I., Sendon, P.M.D., Sanchez, D.L., Ismail, A.M., Mackill, D.J. (2013). QTL mapping and confirmation for tolerance of anaerobic conditions during germination derived from the rice landrace Ma-Zhan Red, *Theor. Appl. Genet*. **126(5)**:1357–1366.

Septiningsih, E.M., Pamplona, A.M., Sanchez, D.L., Maghirang-Rodriguez, R., Neeraja, C.N., Vergara, G.V., Heuer, S., Ismail, A.M., Mackill, D.J. (2009). Development of submergence-tolerant rice cultivars: the Sub1 gene and beyond. *Ann. Bot.* **103**: 151–160.

Shobarani, N., Prasad,G.S.V. *et al.* (2010). *Rice alamanac- India*, DRR Technical bulletin No. 50, Directorate of Rice Research, Hyderabad.

Singh K.N. (1994). Crops and agronomic management. In: Rao E.A., ed. Salinity management for sustainable agriculture, 25 years of research at CSSRI. Karnal, India: Central Soil Salinity Research Institute, pp. 124–144.

Singh,R., Yashi Singh, Kathiresan, R.M., Paramsivam, K., Nadarajan, S., Thirumeni, S., Nagarajan,M., Singh, A.K., Prashant Vikram, Arvind Kumar, Septiningshih, U.S. Singh, Ismail, A.M., Mackill,D., Nagendra K., Singh, (2016). From QTL to variety-harnessing the benefits of QTLs for drought, floodand salt tolerance in mega rice varieties of India through amulti-institutional network. *Plant Science* **242**: 278–287.

Swaminathan,M.S. (2007). Can science and technology feed the world in 2025? *Field Crop Research.* **104**:3-9.

Thomson, M.J., Ismail, A.M., McCouch, S.R., Mackill, D.J. (2010). Marker assisted breeding. In: Pareek, A., Sopory, S.K., Bohnert, H.J., Govindjee (Eds.), *Abiotic Stress Adaptation in Plants: Physiological, Molecular and Genomic Foundation.* Springer, Dordrecht, pp. 451–469.

Xu, K., Mackill, D.J. (1996). A major locus for submergence tolerance mapped on rice chromosome 9. *Mol. Breed.* **2**: 219–224.

Xu, K., Xia, X., Fukao, T., Canlas, P., Maghirang-Rodriguez, R., Heuer, S., Ismail, A.M., Bailey-Serres, J., Ronald, P.C., Mackill, D. J.(2006). Sub1A is an ethylene response factor-like gene that confers submergence tolerance to rice. Nature, **442**: 705–708.

Xu, K., Xu, X., Ronald, P.C., Mackill, D.J. (2000). A high-resolution linkage map in the vicinity of the rice submergence tolerance locus Sub1. *Mol. Gen. Genet.* **263**: 681–689.

4

New Prediction Models of Innovative Technologies for Maximising the Production of Greengram and Blackgram in Coastal Areas

M. Pandiyan and P. Sivakumar
Agricultural College and Research Institute
Tamil Nadu Agricultural University
Eachankottai, Thanjavur, Tamil Nadu- 614 902

Introduction

Pulses belong to the taxonomic family Fabaceae, containing over 18,000 species divided into the three sub-families Mimosoideae, Caesalpinoideae and Papilionoideae. Pulses are commonly cultivated for several decades in all over the world because of the nutritional value of their seeds. Among different pulses, soybean, chickpea, common bean, cowpea, green gram, black gram and pigeon pea contribute significantly to serve the major protein source for the diets of large numbers of people living in Asia, Africa, and South America.

Production and Productivity of Pulses in India

In India, the production of pulses has not been able to keep pace with their domestic demand, resulting in import of 4–5 million tonnes of pulses per annum, especially from the countries like Canada, Myanmar and Australia to meet its domestic requirement (*Kumar et al.*, 2018). India is the largest producer (around 25% of global production), however it consumes 27% and imports around 14% of its pulses requirements. The yield of pulses in India is quite low at 781 kg/ha which might be due to various factors. Pulse crops does not show any significant increase in area and production during 1950–51 to 2009–10, however, significant growth in area and production has been recorded during the last five years (i.e., 2010–2011 to 2016–17), with the adoption of high yielding varieties, increased usage of agricultural inputs like fertilizers and manures etc. The pulses under irrigation are cultivated in about 37% of the area while 63% of pulses are grown under rainfed conditions.

The productivity of pulses has increased about 77% at 779 kg/ha during 2016–17 from the level of 441 kg/ha during 1950-51. It is imperative to mention that the New Agriculture Technology (NAT) introduced during mid-sixties has increased the production of food-grains from 50.82 million tonnes during 1950–51 to 275.68 million tonnes during 2016–17 with the increase in area from 97.32 million hectares to 128 million hectares. In order to achieve self-sufficiency in pulses, the projected requirement by the year 2025 is estimated at 27.5 MT and to meet this requirement, the productivity needs to be enhanced to 1000 kg/ha.

Production and Productivity of Pulses in Tamil Nadu

Among the different legumes grown in India, black gram and green gram are the most commonly cultivated pulses in southern part of India, particularly Tamil Nadu for various food purposes. In India, the area occupied by mung bean is about 3.0 million ha with total production of 1.1 million tones but average productivity is 3.20 (q/ha) and and in Tamil Nadu it is 1.1 million ha with total production 0.55 lakh tonnes. In India, black gram is cultivated in an area of about 3.24 million hectares and producing 1.46 million tonnes. Productivity is only 526 kg/ha. In Tamil Nadu, black gram covers an area of about 3.06 lakh hectares with production of 1.70 million tonnes and productivity of 555 kg/ha. The potential yield of black gram and green gram is low due to several reasons viz., cultivation of low yielding varieties, asynchronous maturity and cultivation in the area of problem soils and marginal lands mostly as rainfed crops and with poor management practices.

Pulses are cultivated under rice fallow conditions in about 2.6 lakh hectares in Tamil Nadu which is 30.75% of the total area under pulses in this state. Rice fallow pulses contribute about 40.5% of the total pulse production.

Importance of Pulses

Pulses constitute an essential part of the Indian diet for nutritional security and environmental sustainability. Pulses are the cheapest source of proteins and Indians fulfil 20 to 30 per cent of their protein requirement from pulses, which are also rich in calcium and iron also. Per capita net availability of pulses in India, however, has reduced from 69.0 gm/day to 47.2 gm/day as against WHO recommendation of 80 gm/day.

Evolution of Pulses in India

The evolution of pulses in India is classified into 5 categories *viz.* pre-green revolution period (1960–1970), green revolution (1970–1980), post-green revolution period (1980–1990), post liberalization period (1990–2000) and a period following the trade-spike from 2000–2010. During the green revolution period, the reduction in the variability of paddy and wheat yield coupled with unabated risks in pulses could have led to the substitution of area away from pulses. The post-liberalization period witnessed favourable terms of trade for agriculture over industry and this has potential to change and impact cropping pattern. Finally, the importance of post trade spike period is indicated by the fact that during this period, pulses import increased by as much as 36 per cent.

Important Reasons for Cultivation of Pulses

Pulses in general are nutritionally enriched as they have high protein content, relative to staple cereals. In addition to their nutritional content, there are several reasons that strongly support legume cultivation and adoption. Important reasons for their cultivation include:

1. Pulses are rich in proteins and found to be main source of protein to vegetarian people of India.
2. It is the second important constituent of Indian diet after cereals.
3. They can be grown on all types of soil and climatic conditions.
4. They give ready cash to farmers
5. Pulses being legumes fix atmospheric nitrogen into the soil.
6. They play important role in crop rotation, mixed and intercropping, as they help maintaining the soil fertility.
7. They add organic matter into the soil in the form of leaf mould.
8. Pulses are generally not manured or require less manuring.
9. They are helpful for checking the soil erosion as they have more leafy growth and close spacing.
10. They supply additional fodder for cattle.
11. Some pulses are turned into soil as green manure crops
12. Majority of the pulses crops are short durational so that second crop may be taken on same land in a year.
13. They provide raw material to various industries. *Example*, Dhal industry, Roasted grain industry, papad industry etc.

Constraints in Pulse Production in India

The pulses production has virtually stagnated over the last 40 years. There are mainly two reasons for this. Firstly, 87% of the area under pulses is rainfed. The second reason is that pulses are mainly grown as a residual crop on marginal lands, after diverting the better-irrigated lands for higher yield-higher input crops like cereals and oilseeds.

The low priority accorded to pulse crops may be related to their relatively low status in the cropping system. As a crop of secondary importance, in many of these systems, pulse crops do not attract much of the farmer's crop management attention.

1. The non availability of high yielding varieties with desirable characters suitable for different condition.
2. The non-availability of seeds of high-yielding varieties in the desired quantities is perhaps one of the major constraints in the expansion of pulses. Although more than 200 improved varieties of pulses have been released since 1970's, its impact hardly gets reflected in the yield . The rate of growth of yield of pulses was 0.03 percent over the past four decades.
3. In pulses, there are a number of diseases and insect pests which cause heavy losses resulting in poor production. Though several resistant/ tolerant varieties had been developed by research institutions, the spread of such varieties in the farmer's fields is very limited.
4. Lower production (as compared to demand) and lower stocks in both domestic and global markets have led to a steep rise in prices of pulses.
5. Farmers are not motivated to grow pulses because of yield and price risk probably due to lack of effective procurement.
6. Lack of Technical knowledge of farmers.
7. Grown as rainfed crop.
8. Lack of incentives.

Constraints of Pulse Production in Coastal Regions of Tamil Nadu

The farmers in coastal areas normally cultivate rice, groundnut, pulses (green gram and black gram), minor millets and vegetables for their livelihood purpose. The main reasons for not cultivating pulses in coastal areas are

- Extension of salt affected and drought affected areas
- Not aware about good pulse crops suitable for coastal region

- Not known about yield potential of blackgram by farmers
- Frequent occurrence of flooding during rainy and cyclone
- Out breaking of pest and diseases
- Believe on zero tillage.

Possibilities for Pulse Crops Cultivation in Coastal Region of Tamil Nadu

1. The Coastline of Tamil Nadu is located on the southeast coast of Indian Peninsula, and forms a part of Coromandel Coast of Bay of Bengal and Indian Ocean. It is 1,076 km long and is the second-longest coastline in the country after Gujarat. **Inadequate rains and droughts have been leading to groundwater depletion in many coastal villages in Tamil Nadu.** This has led to seawater seeping in and increasing the salinity levels in water and soil.
2. The cultivation of black gram and green gram in coastal areas of Tamil Nadu is actually increasing the acreages of pulse production and getting additional income for farmers from these pulse crops in short period of time. In general, most of the tropical grain legumes are highly sensitive to salinity and alkalinity problems, but green gram is cultivated to some extent in inland and coastal salinity area.
3. The evolution and development ideal genotypes are normally tedious and slow. Such plant type evolutes itself. It becomes viable with various morphological, physiological and biochemical characters involved or combined in a single genotype from different sources exhibiting remarkable potential in future. In this case research priority should be given for development of ideal plant type through plant breeding. Ideotype variety is having their own limited capacity for economic yield (20–30 %) without any support, while where other technologies are used, the ideotype will give more yield than our expectation.

Factors and/or Traits of Importance for Enhancement of Pulse Production

- Developed plant types should have shorter duration, more number of productive pods and seed weight, with increased level of per day productivity.
- Pulse ideotype requirement for irrigated condition must be medium stature semi-erect and compact, responsive to high input and with high HI.
- The following are characters essential for developing ideal plant types, varieties and or hybrids in pulses.

General Plant Ideotype Characteristics in Pulses

- Determinate plant type
- Erect and upright plant
- Average plant height
- Early vigour, early flowering and synchronous maturity
- Pod bearing from well above the soil surface
- More pods/plant and more number of seeds/pod
- High harvest index
- Yield stability.

Prediction Model for Enhancing the Pulse Production

- Developing efficient plant architecture
  - The important strategy for increasing productivity of pulse crops has to be achieved based on ideal plant architecture having high biological yield with optimum harvest index.
  - The upright growth habit reduces foliar disease incidence in the canopy by allowing greater airflow and has improved harvest ease.
  - More number of leaves, small, erect, leaf lets should be lanceolated with pale green colour
- Developing up right branched type (up right) and mono stem genotypes
- Developing genotypes with continuous multi blooming and elongated internodes length
- Exploitation of wild species for developing new plant types towards higher yield

  Example: Transfer of three peduncle/axial character from *V. umbellata* to green gram and black gram genotypes
- Evolving genotypes with arrangement of pods in top portion of the plant
  - Development of long duration, perennial type genotypes
  - Developing drought and salinity tolerant genotypes
  - Possibilities of exploitation of heteorsis through CMS and GMS
- Developing genotype with fast growing nature at early vegetative growth stage to surpassing the height of paddy stubbles to avoid etiolating and get light harvesting capacity more in rice follow condition
- Pod bearing starting from 15 cm above the root surface of the plant
- Developing black gram genotype with 8-10 seeds/pod and 15 seeds/pod in green gram

- Developing genotypes with uniform vegetative growth
- Clear distinction between vegetative and reproductive phase
- Developing plants with reduced flower drop
- Developing plants with halophyte root system and long tap root system to withstand salinity
- Inducing shallow anchoring side roots for increased uptake of top soil nutrients
- Enriching bio-fortification genotypes in pulses especially Zn
- Developing of seed priming methods to avoid early drought stress
- ▪ Seed priming, its simplicity and no requirement for expensive equipment and chemical, could be used as a simple method for overcoming a poor germination and seedling establishment in saline condition.
- ▪ The development of suitable priming agents is highly helpful for pulse genotypes to escape from seedling terminal stress when the genotypes are grown in in salt affected soils
- Change of plant biochemical system similar to cereals

  Example: determination of vegetative and reproductive phase
- Erect and upright plants with average plant height
- Early vigour
- Determinate growth habit with synchronous and early maturity
- More pods/plant and more number of seeds/pod
- Resistance to shattering and sprouting.
- Developing genotypes suitable for machine harvest
- ▪ Availability of cultivars suited to mechanical harvesting will reduce production cost and attract farmers towards increased pulses cultivation.
- Higher harvest index
- Lenthy pods to contain more number of seeds
- Leaf shedding during harvest is better.

Merits of New Plant Breeding

1. New plant breeding is an effective method of enhancing yield through manipulation of various morphological and physiological crop characters. Thus, it exploits both morphological and physiological variation.
2. In this method, various morphological and physiological traits are specified and each character or trait contributes towards enhanced yield.

3. New plant type breeding involves experts from the discipline of plant breeding, physiology, biochemistry, entomology and plant pathology. Each specialist contributes in the development of model plants for traits related to his field.
4. New plant type breeding is an effective method of breaking yield barriers through the use of genetically controlled physiological variation for various characters contributing towards higher yield.
5. New plant type breeding provides solution to several problems at a time like disease, insect and lodging resistance, maturity duration, yield and quality by combining desirable genes for these traits from different sources into a single genotype.
6. It is an efficient method of developing cultivars for specific or environment.

Demerits of New Plant Type Breeding

1. Incorporation of several desirable morphological, physiological and disease resistance traits from different sources into a single genotype is a difficult task. Sometimes, combining of some characters is not possible due to tight linkage between desirable and undesirable characters. Presence of such linkage hinders the progress of new plant type breeding.
2. New plant type breeding is a slow method of cultivar development, because combining together of various morphological and physiological features from different sources takes more time than traditional breeding where improvement is made in yield and one or two other characters.
3. New plant type breeding is not a substitute for traditional or conventional breeding. It is a supplement to the farmer. New plant type is a moving object which changes with change in knowledge, new requirements, national policy, etc. Thus new plant types have to be evolved to meet the changing and increasing demands of economic products.

References

Choudhary, AK. and Vijayakumar, AG. (2012). Glossary of Plant Breeding, A Perspective. LAP LAMBERT Academic Publishing, Germany.

Kumar, K., Bhatia J. and Kumar, M. (2018). Constraints in the production and marketing of pulses in Haryana. Int. J. Pure App. Biosci., **6(2)**: 1309-1313.

5

Scaling up of Traditional Paddy Varieties: A Tool to Combat Climate Change

A. V. Balasubramanian, R. Manikandan, A. Rajesh and Subhashini Sridhar
Centre for Indian Knowledge System, Sirkazhi, Tamil Nadu, India

Introduction

Rice is known as the grain of life, and is synonymous with food for Asians. In addition to being a staple food and an integral part of social rites, rituals, and festivals in all Asian countries, it has a medicinal value too, which was clearly recognized by the medicine systems of the region. Ancient Ayurvedic treatises laud the Raktashali red rice as a nutritive food and medicine. The medicinal value of other rices such as Sashtika, Sali, and parched rice have been documented in the Charaka Samhita (c. 700 BC) and the Susruta Samhita (c. 400 BC), in the treatment of various ailments such as diarrhea, vomiting, fever, hemorrhage, chest pain, wounds, and burns.

According to late Dr. Richaria, the well known rice scientist, 4, 00,000 varieties of rice existed in India during the Vedic period. According to his estimates, even today 2, 00,000 varieties of rice exist in India – a truly phenomenal number. Farmers in every part of country have a deep knowledge of their own rice varieties, This has enabled them to harvest a crop even under the most severe stress situations. Traditional paddy varieties are location specific and has its own advantages viz. saline tolerant, drought tolerant, pest tolerant, used for specific purpose such as roofing material for houses etc.

Centre for Indian Knowledge Systems is an organization devoted to exploring and developing the contemporary relevance and application of Indian Knowledge Systems focusing in the area of indigenous agriculture. Our focus areas are biodiversity conservation, organic agriculture and Vrkshayurveda (traditional Indian plant science).

Activities

- Survey, collection and documentation of agronomical characteristics of traditional paddy varieties.
- Setting up farmers' community seed banks.
- Evaluation, characterization and multiplication of these varieties.
- Production of quality seeds and distribution.
- Marketing of Traditional paddy varieties through FPC (Farmer producer companies).

Currently, the centre has been conserving 130 paddy varieties in farmers fields and experimental farms.

Why traditional paddy varieties

The traditional paddy varieties possess the dual characteristics - drought and flood tolerant properties. These varieties can be cultivated in saline soils. Due to this unique feature, selected varieties are cultivated every year in coastal areas of Tamilnadu viz. Vedharanyam taluk, Nagapattinam district and Ramanathapuram district. The traditional paddy varieties can be cultivated without chemical fertilizers, weedicide and pesticides application. These varieties do possess medicinal properties, hence fetches good market value compared to conventional varieties

Examples

Varieties Resistant to Flooding - Koomvazhai, Samba mosanam

Drought Resistant Varieties - Vadan Samba, Jil Jil Vaigunda

Resistant to BPH and Caseworm - Sivappu Kuruvikkar

Best Suited for Puffed Rice -Sembalai

Variety Best Suited for Biriyani - Kitchili Samba

Variety Best Suited for Puttu – Pitchavari.

Characteristic features of some traditional varieties.

1. Kuzhiyadichan.

Season: Samba; Crop duration : 105–110 days

Special Features: Suitable for saline soils in lands with good drainage facility. Resistant to drought, pests and diseases. It is also called 'Kulikulichan'. Ideal for lactating mothers, since it increases the milk flow.

2. Kalarpalai

Crop duration: 105–110 days

Special Features: Suitable for saline soils in lands with good drainage facility. Resistant to drought, conditions. Highly suited for "Aval" preparation.

3. Kalanamak

This is the most important scented rice variety of India and derives its name from its black husk. (Kala means Black, Namak means Salt). It surpasses Basmati rice and is considered the finest quality rice in international trade. It is found to perform well in saline soils.

We have conducted trials with the above traditional paddy varieties post tsunami in Sirkazhi taluk, Nagapattinam district and tested the saline tolerant properties of the traditional paddy varieties during 2005–2006.

4. Karunkuruvai for filariasis

One of the traditional paddy varieties called Karunkuruvai is used in the treatment of Filariasis. It is actively in use by Siddha physicians of Tamilnadu and a reference of it has also been made in an ancient Tamil Siddha medical text.

Two of the traditional paddy varieties namely *NeelanSamba* and *Kuzhiadichan* are said to be good for lactating mothers "galactogogues".

Physico chemical properties of traditional paddy varieties

CIKS in collaboration with Department of Nutrition, Ethiraj college, Chennai has conducted study about the nutrition content of selected traditional paddy varieties during 2014–2015. The study was undertaken to analyze the physicochemical properties, nutrient analysis, evaluate the acceptability and to assess the glycemic index of traditional rice varieties. The blood glucose values were measured using glucometer and the glycemic index of each rice variety was calculated using IAUC calculation in non-diabetic subjects

Case Study

It's a double blind study design. The samples were coded from CIKS-01 to CIKS 10. Sample one (CIKS -01) as the control which is white ponni. The confidential and sealed cover having the details of coded samples was given. Sensory study design was employed to evaluate the quality of traditional rice varieties viz. taste, colour, flavour, texture and appearance. Pre test and post test experimental design with control group was employed to check the glycemic index of rice. The fasting and post prandial (1 hr and 2 hrs) blood glucose was measured.

Findings

- Among the samples analysed mappilai samba had high carbohydrate content of about 80g and fiber content of about 7.07g, which is about 3.5 times higher compared to all other samples
- Karungkuruvai had the highest iron content of about 19.63mg which is more than four times than the iron content present in white ponni (4.25mg). It proves to have low glycemic index compared to other traditional paddy varieties .
- Kalanamak has the highest potassium content of 98.82mg.
- Neelam samba had the maximum amount of 80.63mg of calcium.

Name of the variety	Mean GI (With white ponni as control)	Mean GI (With glucose as control)
Karungkuruvai	68.54	53.81
Mappilai Samba	78.02	68.84
Kudhaivazhai	80.90	66.34
Kalanamak	95.45	50.71
Perungkar	97.93	75.84
Kavuni	75.45	52.36
Kullakar	69.44	52.25
Neelam Samba	79.44	84.37
White Ponni	100	56.57

Conclusion: The selected organically grown traditional hand pound raw rice like Karungkuruvai, Mappilai samba, Kudhaivazhai, Kalanamak, Perungkar, Kovuni, Kullakar and Neelam samba had a significant benefits in terms of nutrients and glycemic index. The potential pigments present contribute to the antioxidant property and the peculiar flavour adds value to the highly acceptable product.

Details about cultivation of traditional paddy varieties in Nagapattinam district (samba season 2018–started during Sept 2018)

CIKS has been working with 2000 farmers spread over six blocks in Nagapattinam district. The following table illustrates the scale of traditional paddy cultivation taken up in Nagapattinam district.

Name of the variety	No of farmers	Area (acres)	Production expected (tonnes)
Soorankuruvai	620	450	350
Kuzhiyadichan	540	400	300
Mappillai samba	70	120	90
Kudaivazhai	135	245	180
Karuppu kavuni	27	52	35
Kitchili samba	15	23	27
Seeraga samba	6	19	20
Total	1413	1309	1002

Market support for traditional paddy varieties

CIKS has been involved in promotion and strengthening of farmer producer organizations in collaboration with SFAC,Chennai and NABARD in nine districts of Tamilnadu. Two important FPC's established were viz. Marutham Sustainable Agriculture Producer Company, Thiruvannamalai district and Valanadu Sustainable Agriculture Producer Company, Nagapattinam district. Valanadu company has supplied quality traditional paddy seeds of selected varieties viz. Seeraga samba, Mappillai samba, Kavuni, Karunkuruvai, Kullakar and Soorankuruvai cumulated as 2600 kg seeds to 900 acres spread in four districts viz. Nagapattinam, Cuddalore, Salem and Thiruvarur districts of Tamilnadu. These seeds were produced as per seed standard specification and marketed as TFL (Truthfully labeled seeds). Marutham company has procured 60 tonnes of seeraga samba paddy and supplied rice to consumers of Chennai. Apart from FPC, Sempulam Sustainable Solutions, a private company supported by CIKS has been involved in market promotion of traditional paddy varieties since 2018.

Challenges: Notification for traditional paddy variety seeds, awareness among consumers, brand for traditional paddy varieties are challenges faced under field condition.

Conclusion: Traditional paddy varieties possess all characteristics to combat climate change factors and have good market potential compared to conventional paddy varieties.

6

Physiological and Biochemical Traits Associated With Salinity Tolerance in Crop Plants

P. Boominathan and M. Pandiyan
Agricultural college and Research Institute, Eachangkottai, Thanjavur-614902.

Plants are frequently subjected to environmental stresses such as water deficit, freezing, heat and salt stress. Soil salinity is a major problem in arid and semi-arid regions, where rainfall is insufficient to leach salt and excess of sodium ion down and out of root zone. In Asia, around 21.5 million ha of cultivable land was affected by salinity. India had 8.6 million ha of salinity affected cultivable land. It is observed that the area of salinity has been increasing in India. In Tamil Nadu, the area affected by salinity is around 6 lakh ha. Salinity affected areas are Chengalpattu, Salem, Thanjavur, Trichy, Tirunelveli, Dharmapuri and Ramanathapuram. Salinity is the most serious threats to the agriculture especially in the arid and semi-arid region. The presence of salts in the soil and their effects on the plant physiological mechanisms was major restrictive factor for agricultural productivity. Investigations on such soils have revealed that they were charged with high and abnormal soluble salt concentrations that lowered agricultural yield, limited crop distribution and even lead to extinction of certain species.

Salinity adversely affects plant growth and development, hindering seed germination, seedling growth, enzyme activity, DNA, RNA, protein synthesis and mitosis. There has been a variation in the response of plants to salinity with its growth stages according to quantity and period of exposure to salt. Salinity by competing in nutrition element absorption causes growth reduction. Primary salt injuries include metabolic disturbance and inhibition of growth and development. Secondary salt effects include nutrient deficiency and osmotic dehydration. Excessive amounts of salts, especially sodium chloride (NaCl), in the soil induce osmotic effects, leading to changes in plant metabolism. General symptoms of damage by salt stress are growth inhibition, accelerated development and senescence and death during prolonged exposure. Salt stress causes the reduction of rice yield and severe salt stress may even threaten survival.

Salinity adversely affects the development and growth of rice plant at all the stages of plant life. Young seedlings were more susceptible to salt stress. Kawasaki *et al.* (2001) reported that tillering is an important agronomic trait for grain production in rice and salinity highly affects the tillering in the plants. Salinity stress mainly affected booting and panicle initiation stages implicated greater adverse effect than that of later growth stages (Asch *et al.,* 2000). Salt stress defense mechanisms, including ion regulation, compartmentalization and production of antioxidants, plant hormones and osmoregulation are well established in the salt tolerant plants (Das *et al.*, 2005). It is thought that the repressive effect of salinity on seed germination and plant growth could be related to a decline in endogenous levels of phytohormones.

Osmotic effects during salinity stress

Osmotic adjustment refers to lowering of the osmotic potential due to the net accumulation of solutes in response to salinity stress. Osmotic adjustment was important mechanism in salinity tolerance, because it enabled continuation of cell, stomatal and photosynthetic adjustments, better plant growth and yield production. The compounds mainly involved in osmotic adjustment are the soluble sugars, organic acids, free amino acids, potassium and chloride ions. The nontoxic compatible organic solutes accumulated in the cytoplasmic compartment of cells and inorganic ions toxic to metabolic processes were restricted to the vacuolar compartment. Salinity reduces the growth of the plant through osmotic effects, thus reduce the ability of plants to water uptake. If excessive amount of salt enters the plant, the concentration of salt will eventually rise to a toxic level in older transpiring leaves causing premature senescence and reduced the photosynthetic leaf area of a plant to a level that cannot sustain growth. The outcome of these effects may cause membrane damage, nutrient imbalance, altered levels of growth regulators, enzymatic inhibition and metabolic disfunction, including photosynthesis which ultimately leads to plant death (Hasanuzzaman *et al.*, 2012).

Nutrient imbalances

Plant growth and development is expressed in terms of crop yield or plant biomass, which might be adversely affected by salinity, induced nutritional disorders. Salinity caused nutrient imbalances in the plant in general, which in turn caused a decrease in the content of nitrogen, phosphorous and potassium. Salinity reduced phosphate uptake and accumulation in crops grown in soils, whereas in solution cultures, reductions might be due to a competitive process.

Salinity dominated by Na^+ salts not only reduced calcium availability but also reduced its transport and mobility to growing regions of the plant, there by affects the quality of both vegetative and reproductive organs. Reduced plant growth caused by salinity might be attributed to the disturbance in the nutrients, resulting from the decreased uptake of potassium, calcium, magnesium, phosphorous and nitrogen. Numerous reports indicated that salinity reduces nutrient uptake and accumulation of nutrients into the plants. Micronutrient deficiencies are very common under salt stress because of high pH. Flowers and Flowers (2005) reported that salinity has three potential effects on plants including lowering the water potential, direct toxicity of Na and Cl, absorbed and interference with the uptake of essential nutrients and the crop responses to salinity varies with growth stages, concentration and duration of exposure to salt. Excess accumulation of salts mainly Na^+, can cause reduction in agricultural crop yields because most crops are susceptible to high salt concentrations in soil (Munns and Tester, 2008).

Compatible osmolytes

Salt-stress induced synthesis of low molecular weight metabolites. They are also engaged as osmo-protectants in maintaining protein function by protecting them against salt-induced damages. Salt-stress causes numerous metabolic changes in plant. This compatible osmolytes are carbohydrates, sucrose, fructans used to mitigation of salt stress in plants (Burg *et al.*, 2008).

Phytohormones

Phytohormones play an important role in alleviating salt stress and helps in creating the adaption mechanism. Foliar application of phytohormones generally stimulated the accumulation of carbohydrates in plants during salt stress. Plant growth regulators are being widely used to counteract the deleterious effects of adverse environmental stresses on plants.

Cytokinin

Rice tillers develop from tiller buds and cytokinin (CTKs) play an important role in regulating tiller bud growth under salinity conditions. The exogenous application of CTK through seed pre-treatment is involved in plant tolerance to salt stress. Pretreatment with different CTKs altered leaf free polyamines under saline conditions. Kinetin is one of the CTKs known to significantly improve the growth of crop plants grown under salinity. Seed treatment with CTKs is reported to increase plant salt tolerance. Kinetin acts as a direct free radical scavenger or it may involve in antioxidative mechanism related to the protection of purine breakdown.

Brassinosteriods

Brassinolide (BL) is natural plant growth promoting substance that exerts anti-stress effects on plants. The effect of brassinosteriod "24-epibrassinolide" (EBL) on growth, yield and physiological traits of salinity-tolerance in rice was significantly improved. BR regulated stress response because of a complex sequence of biochemical reactions such as activation or suppression of key enzymatic reactions, induction of protein synthesis and the production of various chemical defense compounds. Foliar application of BRs also significantly increased the concentration and total uptake of macro and micronutrients in straw and grains.

Salicylic acid

Salicylic acid (SA) is an endogenous growth regulator of phenolic nature, which participates in the regulation of physiological processes in plants and also provides protection against salinity stress. SA considered as a hormone-like substance, plays an important role in the regulation of plant growth and development, such as seed germination, flowering under salinity conditions. The role of SA in defense mechanism is to alleviate salt stress in plants.

Influence of salinity on morphological parameters

Wide range of variation in stress responses to salinity was found among the cultivars and seedling emergence and early seedling growth stages were most sensitive to salinity. Higher level of salt stress inhibits the germination of seeds while lower level of salinity induces a state of dormancy. Seed treatment with kinetin and GA_3 reduce inhibitory effects of NaCl levels during germination stages. Application of BR increases tolerance in rice during germination under salinity stress during germination. The reduced growth and photosynthesis are the main effects of salt stress. The relation between sodium concentration in plant tissue and growth were observed to be negative, and with greater effects on shoot growth than root growth.

The plant height, number of branches, length of branches and diameter of shoot were significantly affected by salt stress. NaCl stress reduced plant height, this reduction of plant height increased gradually with increasing of NaCl concentration. GA_3 application increase the height significantly under salinity conditions. NaCl and SA applications had a significant effect on plant height. Adverse effect of salinity on the root length might be due to lower uptakes of water and nutrients from the growing media due to higher concentration of salts present in the root zone, which may causes imbalances in osmotic pressure. Reduced growth under salt stress might be due to reduced transport of essential nutrient to the shoot. Leaf area has been shown to be highly correlated to grain yield in rice under salt stress. Application of GA_3 significantly increased leaf area under salinity stress in rice. Leaf area was negatively affected with salinity stress, while SA treatment resulted in increased leaf area (Ismail *et al.*, 2013). Leaf area was highly sensitive to salt with about 50 per cent reduction even at the lowest concentration. Under salt stress, consumption of metabolic energy increased while the amount of carbohydrate accumulation decreased in salinity conditions. The stress suppressed plant height, leaf number and size, and tillers, which finally lowered the dry matter. The fall in the dry weight was in the elevated metabolic energy cost and reduced carbon gain due to salinity.

Influence of salinity on physiological parameters

Leaf water potential

Soluble sugars play key role in osmoregulation including controlling of leaf water potential and osmotic potential in plant cells under saline conditions. Increase of salt in the root medium lead to a decrease in leaf water potential and hence, may affect many plant processes. High salinity stress also delays the emergence of nodal roots, leaf and tiller with decrease in relative growth rate (RGR), leaf area ratio (LAR) and specific leaf area (SLA).

Relative water content

Increased salinity level caused reduction in leaf relative water content (RWC) content in the plants. A decrease in relative water content indicates a loss of turgor that results in limited water availability for cell extension processes. Foliar application of kinetin elevated relative water content in salt-stressed plants. Similarly, it has been reported that kinetin treatment improved the water status of rice plants grown in high salinity stress.

Photosynthesis

The reduction in the photosynthetic ability was observed in response to salt induced osmotic stresses in rice. Several physiological pathways like photosynthesis, respiration, nitrogen fixation and carbohydrate metabolism have been affected by high salinity conditions. Under saline conditions, decrease in levels of chlorophyll content and stomatal conductance was observed which ultimately reduced photosynthetic rate.

Sugar is a primary product of photosynthesis in the leaf tissues (source) and then transferred to other organs (sink) using phloem loading, which is negatively affected by salt stress, resulting in low sucrose levels in the root tissues (Lemoine *et al.*, 2013). Soluble sugars have been suggested to play a major role in osmotic adjustment in the cellular level in salt defense mechanism in salt tolerant plants. Salt stress can lead to stomatal closure, which reduces CO_2 availability in the leaves and inhibits carbon fixation, exposing chloroplasts to excessive excitation energy, which in turn could increase the generation of Reactive Oxygen Species (ROS).Application of brassinosteriods increased photosynthesis by increasing stomatal conductance in salt stressed plants and may have contributed to the enhanced growth (Choudhary *et al.*, 2012).

Transpiration rate

Translocation of salt into roots and to shoots is an outcome of the transpirational flux required to maintain the water status of the plant and unregulated transpiration may cause toxic levels of ion accumulation in the shoot. Lower transpiration rate, coupled with reduced ion uptake by the roots or reduced xylem loading, may cause poor supply *via* xylem.

α-amylase activity

The α-amylase is an important key enzyme in carbohydrate metabolism that plays an important role in starch degradation in germinating seeds and mobilization of reserved material required for the growth of young seedling. α-amylase activity in germinating seed represents an important factor contributing to seedling development and vigor, which is an important agronomic trait under salinity conditions.

Under normal conditions, synthesis of α-amylase occurring during germination leads to changes in the metabolic activity required for seed germination. Salt stress prevents the synthesis of α-amylase and specific metabolic activity and inhibits the germination. Gibberellic acid (GA_3) is well known to induce the synthesis of α-amylase and hydrolysis of starch in rice seeds in normal conditions

Chlorophyll content

Decrease in chlorophyll content under salt stress is a phenomenon, used as a sensitive indicator of the cellular metabolic state in the plants. The effect of salinity on chlorophyll synthesis and integrity seems to vary with the level of salt stress, as an accelerated rate of biosynthesis and higher concentrations during vegetative growth (Asch *et al.*, 2000). It has been reported that the typical symptom of salinity injury to plant is the growth retardation due to the inhibition of cell elongation and the decrease in chlorophyll content under salinity conditions. Salinity leads to dehydration and osmotic stress resulting in stomatal closure, reduced CO_2 supply and a high production of reactive oxygen species causing irreversible cellular damage and photo-inhibition.

Much evidence suggests that cytokinins (CKs) are the major leaf senescence-inhibiting hormones, since senescence is delayed after the exogenous application of CKs. Brassinosteriods remove the inhibitory effect of salt stress on chlorophyll pigment levels and growth stimulation by brass inosteriods under saline conditions (Hasegawa *et al.*, 2000). The application of salicylic acid promoted salinity tolerance in rice and increases the content of chlorophyll and carotenoid and maintained membrane integrity, which was associated with more K^+ and soluble sugar accumulation in the root under saline condition. Chlorophyll stability index (CSI) was higher in response to a combination of NaCl and SA derivatives which are found in the plants very commonly and having hormone-like effect, can decrease the adverse effects of salinity stress.

Proline content

Amino acids represent one of the most important classes of metabolites in the cell due to the fact that they are the building blocks of proteins, which forms the chemical basis necessary for life and have a variety of roles in metabolism (Chinnusamy *et al.,* 2005). Free amino acids, especially proline can accumulate in a variety of species and serve salinity conditions. Proline is one of the important components of defense reactions of plants to salinity. It might be expected that pretreatment with salicylic acid contributes to accumulation of this amino acid under salt stress. Khedr *et al.* (2003) reported that proline induces the expression of salt stress responsive proteins and may improve the plant adaptation to salt-stress. Endogenous proline accumulation in salt stressed plants has been utilized as effective indicator for salt tolerance.

Catalase activity

Antioxidant enzymes such as superoxide dismutase (SOD) and catalase (CAT) have been considered to act as a defensive team, whose combined purpose is to scavenge reactive oxygen species under salt stress conditions. Salinity comprising both osmotic and ionic effects is known to induce secondarily an oxidative stress in plants, forming reactive oxygen species of various natures. The reactive oxygen species (ROS) are highly cytotoxic and if remain unscavenged, can react with vital biomolecules like protein, nucleic acid, lipids etc. Under salt stress condition, high level of catalase enzymatic activity was observed compared to the normal plants. The exogenous application of GA_3 helps the plant to ameliorate the abiotic stress conditions, since it may provide a mechanism to regulate the metabolic process as a function of sugar signalling and anti-oxidative enzymes.

Starch content

Ismail *et al.*, (2007) recorded that decrease of starch content in shoot tissues to the decrease in the amount of K+ absorbed under salt stress. Another reason for reduction in starch concentration in plant tissue is the direct effects of decreased CO_2 assimilation caused by reduction in stomatal conductance and content of chlorophyll in plant tissues under salt stress. Sugar content in shoots had a significant increase under salinity stress but the starch content of roots of seedlings decreased in NaCl stressed seedling.

Carbohydrates

The carbohydrate metabolism is an important physiological response for adaptation to abiotic stress in plants and various metabolic changes in salinity conditions. Soluble carbohydrates and starch, accumulates under normal conditions before the stress commonly constitute the main resources for plants to supply energy during stress condition, as well as during recovery of the plant. Under most salt stresses, the ability of plants to recover from stress normally increase with increasing concentrations of photosynthetic assimilates in plant tissues during or after stress.

Nutrient content

Salinity associated with excess NaCl adversely affects the growth and yield of plants by depressing the uptake of water and minerals and normal metabolism. Essah *et al.* (2003) observed that excess Na^+ in plant cells directly damages membrane systems and organelles, resulting in plant growth reduction and abnormal development prior to plant death. Excessive salts injure plants by disturbing the uptake of water into roots and interfering with the uptake of competitive nutrients. Na^+ accumulation and Na^+/K^+ ratio in salt-stressed plants depend on salt stress treatments, while K^+ accumulation is generally decreased in the plants.

This effect could be mediated by the lower K^+ and the unfavorable ratio of Na^+ to K^+ in plant tissue under salt stress with the consequent effects on the activity of the enzymes involved in the translocation and conversion of soluble sugars into starch (Ismail *et al.*, 2007). The potassium uptake is usually inhibited under salt stress, because of its molecular similarity to sodium ions, causing competition during active uptake. Therefore, high K content or lower Na/K ratio can be considered as one of major tolerance trait in crop plants.

References

Asch, F., M. Dingkuhn and Doerffling, K. (2000). Salinity increases CO_2 assimilation but reduces growth in field-grown irrigated rice. *Plant Soil*, **218**: 1–10.

Burg, M.B., Kwon,E.D.and Kultz, D.(2008). Osmotic regulation of gene expression *FASEB. J.,* **10:** 1598–1606.

Chinnusamy, V., Jagendorf, A. and Zhu, J.K.(2005). Understanding and improving salt tolerance in plants. *Crop Sci.,* **45**: 437–448.

Choudhary, S.P., Yu,J.Q., Yamaguchi-Shinozaki,K., Shinozaki, K. and Lam-Son, P.T. (2012). Benefits of brassinosteroid crosstalk. *Trends Plant Sci.,* **17**: 594–605.

Das, K.K, Sarkar, R.K. and Ismail, A.M. (2005). Elongation ability and nonstructural carbohydrate levels in relation to submergence. *Plant Sci.,* **168**: 131–136.

Essah, P.A., Davenport, R. and Tester, M.(2003). Sodium influx and accumulation in Arabidopsis. *Plant Physiol.,* **133**: 307–318.

Flowers, T.J and Flowers, S.A. (2005). Why does salinity dose such as difficult problem for plant breeders? *Agric. Water Manage.* **72:** 15–24.

Hasanuzzaman, M., Hossain,M.A., da Silva, J.A.T. and Fujita, M. (2012). Plant responses and tolerance to abiotic oxidative stress: antioxidant defenses akey factors. Crop stress and its management: perspectives and strategies. *Springer*, Berlin, pp 261–316.

Hasegawa, P.M., Bressan,R.A., Zhu, J.K. and Bohnert, H.J. (2000). Plant cellular and molecular responses to high salinity. *Ann. Rev. Plant Physiol. Mol. Biol.,* **51**: 463499.

Ismail, A.M.,Heuer, S.,Thomson, M.J. and Wissuwa, M. (2007). Genetic and genomic approaches to develop rice germplasm for problem soils. *Plant Mol. Biol.,* **65**: 547–570.

Ismail, A.M., Singh,U.S., Singh,S., Dar, M.B. and Mackill, D.J. (2013). The contribution of submergence-tolerant. *J. Agric. Res.,* **13:** 705–710.

Kawasaki, S., Borchert,C., Deyholos, M. and Wang, H. (2001). Gene expression profiles during the initial phase of salt stress in rice. *Plant Cell,* **13:** 889-905.

Khedr, A.H.A., Abbas,M.A., Wahid,A.A.A.,Quick, W.P. and Abogadallah, G.M.(2003). Proline induces the expression of salt-stress responsive proteins and may improve the adaptation of *Pancratium Maritimum.* (L.) to salt-stress. *J. Exp. Bot.,* **54**: 2553-62.

Lemoine, R., La Camera,S., Atanassova,R., Dédaldéchamp,F., Allario, T. and Durand, M.(2013). Source-to-sink transport of sugar and regulation by environmental factors. *Front Plant Sci.*, **4:**1–21.

Munns, R. and Tester, M.(2008). Mechanisms of salinity tolerance. *Ann. Rev. Plant. Biol.,***59:** 651–681.

7

Microbiome Component for Sustainable Management of Soil Fertility and Productivity in Coastal Farming

N. Ramanathan and K. Sivakumar
Department of Microbiology Faculty of Agriculture
Annamalai University

Life on Earth is dependent on microorganisms for many essential services. Microbes are intimate partners in global agriculture by way of mobilization of nutrients, antagonism against pathogens and exploitation of plant- microbe interactions and optimization of plant microbiome allows farmers to apply less chemical fertilizers and pesticides, and improving plant growth and yield for sustainable development.

India has 2.4% of land area of world and 8.1% global species diversity. India has 7500 km long coast line in nine states and three union territories. About 14% of Indian population (17 crores) lives in coastal area. In 10.78 million ha of coastal ecosystem, rice is the major coastal crop. Coastal rice accounts for 15% of total area and production of India. Coconut, cashew, areca nut, spices, plantation and fruit crops are the other important coastal crops. The east coast in four states (Tamil nadu, Andhra Pradesh, Orissa, West Bengal) and one union territory (Pondicherry) has sandy coast and plains with medium to high rainfall (100–250 cm), and hot humid climate. It has marshy lands and mangroves. The west coast in five states (Kerala, Karnataka, Goa, Maharashtra, Gujarat) and two union territories (Diu and Daman, Dadar and Nagarhaeli) has fertile sandy plains with high rainfall (250–300 cm) and hot humid climate suitable for horticulture and plantation crops. Western Ghats near west coast is a hot spot with high biodiversity. ICAR established the Central Coastal Agriculture Research Institute (CCARI) at Goa in 2015 to carry out research on coastal agricultural crops.

Soil salinity in coastal regions is not only a soil type character, but also due to low quality irrigation water. The plant associated microorganisms"Plant microbiome" play a significant role in conferring resistance to soil salinity. The endophytic bacteria- rhizobium, rhizobacteria- azospirillum, azotobacter, phosphobacteria, PGPR bacteria, cyanobacteria, symbiotic fungi- ecto and endo mycorrhiza, alleviate the impact of soil salinity. Minimal microbiome for plants is the key determinant of plant health and productivity. The plant-microbe interactions are of three types: (i) **Positive**: Beneficial organisms-nitrogen fixers, phosphobacteria, mycorrhiza; (ii) **Negative**: Harmful organisms - plant pathogens; (iii) **Neutral**: Transient organisms. Millions of microbes inhabit plants, form complex ecological communities to improve plant growth, crop quality and productivity.

Plant microbes help in

- Mobilize nutrients to crops – Biofertilization
- Promote plant growth, PGPR – Biostimulation
- Antagonism against pathogens – Biocontrol
- Plant stress alleviation- to adapt them to salinity – Bioremediation

Biofertilizers for sustainable Agriculture

The present day environmental pollution and its health hazards focused much attention on integrated nutrient management (INM), involving organic and inorganic sources of plant nutrients to sustain the agriculture production. Biofertilizers form an integral part of organic farming as source of nutrients to crop plants. India is one of the largest producers of biofertilizer in the world. In recent years, use of biofertilizers or microbial inoculants to enrich nitrogen and phosphorus nutrients in soil has become a practice in most of the countries as far as economic and environmental viewpoints are concerned. Biofertilizers based on renewable energy sources, are cheaper and environment-friendly microbial fertilizers, among various non-mineral sources of nutrients. It can play a very significant role in improving soil fertility by fixing atmospheric nitrogen, both in association with plant root and without it, solubilize insoluble soil phosphates, mobilize the soil 'P' and produce plant growth substances in the soil.

Biofertilizers are, microbial preparations containing live or latent cells of efficient nitrogen fixing, phosphate solubilizing, phosphate mobilizing, plant growth promoting organisms used for application to seed or soil with the objective of enhancing the availability of nutrients to increase the crop yield. Biofertilizers are highly efficient microbial preparations used for improving the grain yield and soil texture for long term sustainable agriculture.

Advantages of biofertilizers over chemical fertilizers

1. The cost of chemical fertilizers are ever increasing as they are produced from petroleum products and the gap between demand and supply of chemical fertilizers is widening.
2. The self sustainability of soil is lost through continuous application of chemical fertilizers alone.
3. Nutrient utilization efficiency of chemical fertilizers are low as the nutrients are lost through volatalization, leaching and conversion to unavailable forms.
4. The chemical fertilizers cause environmental pollutions- soil, water and air pollution.
5. Biofertilizers are cheap, renewable nutrient sources and possess high nutrient utilization efficiency and eco-friendly.

The biofertilizer usage resulted in increased yield in several crops and resulted in the reduction of cost of cultivation of crops apart from the improvement of soil fertility. Biofertilizers substitute 25 to 30 per cent of nitrogen in a short period and shows increased growth and yield. The ever-increasing cost of the nitrogenous fertilizers emphasized the need for full exploitation of biological nitrogen fixation. In other hand, it reduces the chemical pollution and cost of nutrient input.

Nitrogen is one of the important essential nutrients for plant growth and the most extensively applied plant nutrient. The demand is mostly solved by chemical fertilizers, which would create more environmental problems. Despite greater abundance of nitrogen in the atmosphere, it is in an unavailable form for almost all the living systems, except for a few prokaryotic microorganisms. Biological nitrogen fixation is the most important supplement for the increasing high costs of nitrogen fertilizer input into cropping systems, without substantial loss in yield. The ecofriendly microbial inoculants or biofertilizers offer an opportunity to introduce renewable nutrient sources. Application of biofertilizers improve physical, chemical and biological properties of the soil. Biological nitrogen fixation is one of the beneficial aspects of plant-microbe interaction and gaining greater importance. There are symbiotic, nonsymbiotic and associative symbiotic nitrogen fixing organisms (diazotrophs), which are differentiated based on their interaction with host plants and nitrogen fixing activity.

1. Rhizobium biofertilizer

Rhizobium is a soil bacterium that fixes atmospheric nitrogen in root nodules and lives in symbiotic association with leguminous plants. Rhizobium lives freely in the soil and symbiotically in leguminous plant roots (Singh *et al.*, 1998).

The nodulated legumes contribute a sizable amount of nitrogen fixed in the biosphere. The role of legumes in enriching the fertility of soil was known through centuries. The soil fertility in semiarid tropics is largely maintained by biological nitrogen fixation by legumes. The production of nodules in the roots of leguminous plants by the bacteria of the genus, Rhizobium is one of the most significant microbiological processes in agriculture. The soil bacterium, Rhizobium has the ability to fix atmospheric nitrogen in symbiotic association with legumes in the root nodules. Hence they are also known as 'root bacteria'. The high cost of commercial chemical fertilizer nitrogen calls for an assessment of role of legumes in improving the fertility status of the soil. The nodulated legumes contribute the major amount of nitrogen fixed in the biosphere (Yao *et al.,* 2002). Different legumes are grown in various agro-climatological regions of our country. Leguminous plants not only utilize biologically fixed nitrogen in root nodules for their growth, but also add considerable amount of nitrogen into soil for the benefit of subsequent crop in crop rotations. Rhizobia can enter into symbiosis with leguminous plants by infecting their roots and forming root nodules. The term symbiosis generally denotes a mutual beneficial relationship between two organisms.

Benefits of Rhizobium biofertilizer

1. Rhizobia fix atmospheric nitrogen in the root nodules (about 60 to 130 kg N/ha) and meet 90% of nitrogen requirement of the legumes.
2. Rhizobium application has been found to increase the crop yield by 15–30%.
3. Rhizobium inoculation also leaves sizeable amount of nitrogen in the soil for the benefit of succeeding crop in rotation.
4. Rhizobium forms nodules both on roots and stems in some crops which makes the crop better suited for green manuring.
5. Endophyte Rhizobium is not influenced directly by soil salinity. It helps to maintain leaf relative water content; Selective uptake of K+ ions resulted in salt tolerance and decreased electrolyte leakage; Increases proline production in host plants; Proline synthesized in plants in the presence of rhizobacteria under salinity protects membranes and proteins against adverse effects of salinity.

2. Azospirillum biofertilizer

Some diazotrophs could proliferate better in root zone of crop plants and in the process could exchange metabolites, receive carbon compounds from plant roots and release reduced nitrogen which is absorbed by plant roots due to its proximity (Sahoo *et al.,* 2014). This kind of association is termed as "associative symbiosis" – a term coined to designate association of Azospirillum to cereal root system. Associative symbiosis is intermediate between symbiotic and non-symbiotic associations. The association of Azospirillum with plant roots does not result in the formation of an easily detectable plant structure. Nitrogen fixing bacteria like Azospirillum in the root region are potential suppliers of nitrogen for crop plants (Tejera *et al.*, 2005). Azospirillum substantially contribute to the nitrogen economy of a number of crop plants through biological nitrogen fixation. Agronomic application of Azospirillum over the past 30 years proved that these bacteria are capable of promoting yield of agriculturally important crops in different soils and climatic conditions. Azospirillum species are capable of fixing nitrogen in the range of 40–60 kg/ha/year, which can also improve crop production by hormonal stimulation. Bioactive substances produced by Azospirillum has similar effect as that of growth regulator application.

Azospirillum is a very common soil and root bacterium found in root region (rhizosphere), root surface (rhizoplane) and also inside the root (endorhizosphere). Azospirillum species are frequent inhabitants of rhizosphere of wide variety of crop plants in diverse climatic regions of the world.

Salient features of Azospirillum biofertilizer

Azospirillum species fix atmospheric nitrogen in their cells and associated with the roots of crop plants. It can supply 40 to 60 kg N/ha/year. Azospirillum supplies nitrogen to crop plants at the critical stages i.e., flowering period, when fertilizer nitrogen is depleted. The utilization of biological nitrogen fixation by Azospirillum with cereal crops is extremely interesting due to the area of lands occupied by them and also to the high response of such crops to nitrogen. Hence Azospirillum is the potential biofertilizer for cereal crops. The benefit from Azospirillum is not restricted to nitrogen fixation but to the very complex characteristics such as hormonal effect, improvement of mineral and water uptake. Azospirillum produces growth promoting substances (phyto-hormones) like Indole 3 Acetic Acid (IAA), gibberellins and cytokinins and promotes root proliferation of inoculated plants. Azospirillum inoculation enhanced root branching and root hair density resulting in increased uptake of mineral and water. *A. brasilense* is capable of efficiently colonizing and elongating the root zones. Inoculated plants extract more soil moisture from deep soil layers and withstand stress conditions viz., drought and salt stress. Seed inoculation of Azospirillum increased seed germination, seedling height and vigour index. Azospirillum inhibits phytopathogenic microflora in the rhizosphere by competition and production of iron chelating compounds, siderophores. Azospirillum increases proline content in plants for alleviation of salt stress, which protects membranes and proteins against adverse effects of salinity.

3. Phosphobacteria

The soil bacteria belong to the genera Pseudomonas and Bacillus possess the mechanism to solubilize insoluble phosphate into soluble phosphate, by the ability of producing organic acids such as formic, acetic, propionic, lactic, fumaric, succinic oxalic, gluconic, glycotic and malic acids. *Bacillus megaterium* var phosphaticum, *Bacillus circulans*, *Pseudomonas striata*, *Pseudomonas liquifaciens* are the common soil bacteria that possess the phosphate solubilizing ability (Zou *et al.,* 1992). Phosphate solubilizing microorganisms (PSM) mineralize organic matter to liberate orthophosphate and produce phosphatase, phytase, phospholipase enzymes that release phosphates from organic substrates. They produce (i) carbonic acid, (ii) inorganic acids, (iii) hydrogen sulphide, which release available forms of phosphates to plants. Phosphobacteria improves absorption of phosphorus, which in turn increase growth rate of plants, salt tolerance and suppress adverse effects of salinity (Sharma *et al.,* 2013). Phosphobacteria produce exopolysaccharides, which bind to sodium making it unavailable to plants in saline conditions. Exopolysaccharides, bind to soil particles to form micro-aggregates, increase resistance of plants to water stress and salinity.

4. PGPR

Many soil bacteria are considered as plant growth promoting rhizosphere bacteria (PGPR) as they secrete growth hormones like indole acetic acid (IAA), gibberellic acid, cytokines that promote plant growth by cell division, cell enlargement, flower, fruit and seed formation (Gray and Smith, 2005). PGPR affects plant growth directly by plant growth promoting substances and indirectly by acting against plant pathogens. Fluorescent pseudomonads have emerged as the largest potentially most promising group of PGPR (Khalid *et al.,* 2009). Fluorescent pseudomonads were isolated from rhizosphere of various crops belonging to *Pseudomonas fluorescens.* Besides promoting plant growth, they also produce antibiotics and siderophores which inhibit plant pathogens (Paul and Nair, 2008). Central Coastal Agricultural Research Institute (CCARI), Goa, developed bioformulation of PGPR, *Bacillus methylotrophicus* for rice under salt affected soils of coastal region.

5. Cyanobacteria

The tropical flooded rice field environment provides suitable conditions for the growth and multiplication of blue greenalgae, with respect to their requirements for light, water, high temperature and nutrients. In rice soils, cyanobacteria

(BGA) are the most important nitrogen fixing organisms because of their auto trophic nutrition. The rice ecosystem allow scyano bacteria to function properly, selectively and effectively (Bertocchi *et al.,* 1990). BGA such as Anabaena, Aulosira, Tolypothrix, Calothrix, Plectonema, *Cylindrospermum Ocillatoria* actively fix the atmospheric nitrogen in the soil. In addition to the nitrogen fixation, BGA also release vitamins B_{12}, auxins, etc., which induce the growth of higher plants. Thus they form effective biofertilizers in Agriculture. BGA increase the soil fertility and maintain the proper physical properties of soil. BGA is involved in the change of pH of soil under certain conditions and the specific blue green algal species. Besides nitrogen fixation, BGA can able to produce amino acids, plant growth hormones, vitamins which are essential for plant growth. They are

- Increase the growth of plants by providing nutrients (N, P).
- Add organic matter and improve soil properties to alleviate soil salinity.
- Cyanobacteria exhibit salt tolerance due to presence of exopolysaccharides.
- Mucilaginous sheath imbibe sodium and spare the crop from salt injury.
- Cyanobacteria salt tolerance is exploited for reclamation of saline soils.
- Cyanobacteria, *Nostoc calcicola*is suitable for growing rice in saline and usar soils. It is useful for reclamation of saline, alkali and calcarous soils.
- Cyanobacterial exopolysaccharides improve soil structure and exoenzyme activities.

6. Mycorrhizal Biofertilizer

The symbiotic relationship between fungi and plant roots is termed as mycorrhiza. Mycorrhiza literally means "Fungus root". Perhaps more than 70 per cent of the species of higher plants have this relationship. The association occurs in most plant families, all types of climates and over a broad ecological range. In this mutualistic, symbiotic association, both partners benefit from the interaction. The fungus is supplied with an energy source by the plant and the plant is supplied with minerals and water by the fungus (Miller and Jastrow, 1992). Mycorrhizal associations are so prevalent that the non-mycorrhizal plant is more the exception than the rule. Mycorrhizal plants increase the surface area of root system for better absorption of nutrients from soil especially when soils are deficient in phosphorus. Anatomically mycorrhiza can be divided into ecto and endo mycorrhizae. Ectomycorrhizae are most common among forest and ornamental tree species in the families, Pinaceae, Salicaceae, Betulaceae, Fagaceae and Tiliacaea. The fungal partners in an ectomycorrhiza most frequently belong to Basidiomycetes (Boletus) or Ascomycetes (Eurotium). Fungal sheath develops around the root and part of the mycelium also invades the intercellular space of peripheral cortical root tissue.

In endomycorrhiza, the fungus penetrates the root tissue and lives in an intercellular and intracellular position. Most endomycorrhizal fungi belong to Zygomycetes. Important genera are; *Glomus, Gigaspora, Acaulospora, Sclerocystis, Entrophospora, Scutellospora.* These fungi form what is known as Vesicular-Arbuscular Mycorrhizae (VAM) with plants. These fungi being obligate biotrophs do not grow on synthetic media. Vesicular Arbscular mycorrhizae (VAM) are the most common type of endomycorrhizae. They form vesicles (storage pouches) and arbuscules (absorptive structure) in the plant roots (Bethlenfalvay and Schüepp, 1994). The latter are finely branched structures to help in the transfer of nutrients to plants especially phosphorus. VAM associations are common in millets, oil seeds, tuber crops, vegetables, spices and plantation crops. The beneficial effect of mycorrhiza is of special importance for those plants having coarse and poorly branched root system(e.g. onion), since the external hyphae absorb nutrients from much larger soil volume.

Benefits of mycorrhizae to plants

Various field experiments have shown that the application of VAM fungi increase the crop yield by 6 to 18 per cent. Moreover it is beneficial to plants to many fold.

1. VAM fungi increase longevity of feeder roots, area of rhizosphere and rate of absorption of water and nutrients from soil.
2. They play a key role in selective absorption and mobilization of immobile phosphorus and other minerals (Zn, Cu, K, and S) to plants.
3. VAM fungi reduce plant response to soil stress such as salinity, toxicity associated with heavy metals and mine spoils.
4. They eliminate root pathogens by competition and antagonism, confer disease resistance to plants.
5. VAM fungi increase biomass productivity, growth and yield of crop plants. The growth responses to VA mycorrhizae are caused by the increased supply of phosphorus to the plant via external hyphae. Mycorrhizal inoculation results in increased phosphorus content of seeds and thus they have a substantial effort on germination and seedling vigour.
6. Salinity stress alleviation mechanism by mycorrhiza: They increase the uptake of limiting plant nutrients i.e., P; Improve plant K+ uptake, reduce translocation of Na under salt stress, resulting in high K:Na ratio; Modify negative influence of Na ions on plant growth; They improve photosynthetic efficiency and improve salinity tolerance.

Recent Developments in salinity stress alleviation by microbes

1. ACC deaminase: ACC- Amino cyclopropane carboxylic acid, is a precursor of ethylene, inhibitor of root growth produced by plants in salt stress. Rhizobacteria produce an enzyme, ACC deaminase, which degrade ACC and supply nitrogen and energy. Bacterial hydrolysis of ACC leads to decrease in plant ethylene level, results in increased root growth and tolerate salinity in soil. ACC deaminase producing rhizobacteria mitigate soil salinity by reducing stress ethylene levels. Inoculating ACC deminase bacteria induce longer roots, results in uptake of more water and minerals from soil.
2. Quorum sensing in bacteria: Bacteria in large numbers as a group modify their behaviour through cell to cell communication by secreting signal molecules (chemicals). Individual cells sense how many cells are there in the surrounding, whether there is enough quorum, to act as a group is called Quorum sensing. Rhizobacteria in the presence of 10 nM Homoserine lactone, a regulatory signal, increase the transpiration in plants so that they can obtain diffusion limited minerals, such as P for their growth. Rhizobacterial growth and multiplication effect plant tolerance to soil salinity through, osmoregulation and proline accumulation.
3. Proteomics: Proteomics is the study of characterization of complete set of proteins present in a cell/organism at given time under specific conditions. Proteomics is superior to Genomics. Proteins are the final products and not genes, which are responsible for phenotypic characters. Root protein exudates have a role in crosstalk as the two organisms (plants and microbes) establish known as Protein based communications. This interaction may be compatible (Lectin in legume-rhizobia symbiosis) or incompatible (Root proteins and small molecules have potential to reduce disease incidence).

Plant microbes play significant role in conferring resistance to soil salinity through

- Triggering osmotic response, induction of osmotic protectors.
- Induction of novel genes in plants.
- Production of microbial exopolysaccharides.
- Ion balance- Improve plant K+ uptake, reduced translocation of Na under salt stress, resulting in high K:Na ratio and modified negative influence of Na ions on plant growth is modified.
- Location and environment specific microbial inoculants developed for coastal agriculture improve plant growth, salt tolerance and suppress adverse effects of soil salinity.

References

Bertocchi, C., Navarini.L., Cesaro, A. and Anastasio, M. (1990). Polysaccharides from cyanobacteria. *Carbohydrate polymers*, **12(2):** 127–153.

Bethlenfalvay, G.J. and Schüepp, H. (1994). Arbuscular mycorrhizas and agrosystem stability. In: Gianinazzi, S., Schüepp, H. (Eds.), Impact of Arbuscular Mycorrhizas on Sustainable Agriculture and Natural Ecosystems. Birkhauser Verlag; Basel, Switzerland, pp. 117–131.

Gray,EJ. And Smith, DL. (2005). Intracellular and extracellular PGPR: Commonalities and distinctions in the plant-bacterium signaling processes. *Soil Biol Biochem.*,**37:**395 412.

Khalid, A., Arshad, M., Shaharoona, B. and Mahmood, T. (2009). Plant Growth Promoting Rhizobacteria and Sustainable Agriculture. Microbial Strategies for Crop Improvement, Berlin. Springer; 133–160.

Miller, R.M. and Jastrow, J.D. (1992). The role of mycorrhizal fungi in soil conservation. In: Bethlenfalvay, G.J., Linderman, R.G. (Eds.), Mycorrhizae in Sustainable Agriculture. Agron. Soc. Am. Special Publication, No. 54. Madison, WI, pp. 24–44.

Paul, D. and Nair, S. (2008). Stress adaptations in a Plant Growth Promoting Rhizobacterium (PGPR) with increasing salinity in the coastal agricultural soils. *J Basic Microbiol.***48:** 378–384.

Sahoo, R.K., Ansari, M.W., Pradhan M, Dangar, TK., Mohanty, S. and Tuteja, N. (2014). Phenotypic and molecular characterization of native Azospirillum strains from rice fields to improve crop productivity. *Protoplasna*, **251:** 943–953.

Sharma, SB., Sayyed, R.Z., Trivedi, MH. and Gopi, TA. (2013). Phosphate solubilizing microbes: sustainable approach for managing phosphorus deficiency in agricultural soils. *Springerplu*s, **2**: 587.

Singh, G.V., Rana, N.S. and Ahlawat, I.P.S. (1998). Effect of nitrogen, Rhizobium inoculation and phosphorus on growth and yield of pigeon pea (*Cajanas Cajan*). *Indian J. Agron.*, **43(2):** 358–361.

Tejera, N., Lluch, C., Martínez-Toledo, MV. and Gonzàlez-López, J. (2005). Isolation and characterization of Azotobacter and Azospirillum strains from the sugarcane rhizosphere. *Plant Soil*, **270:** 223–232.

Yao, Z.Y., Kan, FL., Wang, E.T., Wei, GH. and Chen, WX. (2002). Characterization of rhizobia that nodulate legume species of the genus Lespedeza and description of Bradyrhizobium yuanmingense sp. *Int. J. Syst. Evol Microbiop*, Nov. **52:** 2219–2230.

Zou, X., Binkley, D. and Doxtader, KG.(1992). A new method for estimating gross phosphorus mineralization and immobilization *rates in soils. Plant and Soil*, **147:** 243–250.

8

Saline Soil in Coastal Ecosystem Issues and Initiatives

M.V. Sriramachandrasekharan
Department of Soil Science and Agricultural Chemistry
Faculty of Agriculture, Annamalai University

Introduction

The coastal tract of peninsular India extends from Rann of Kutch in Gujarat to Malabar Coast in Kerala on the western coast and from Coromandal coast in Tamil Nadu up to Sunderban delta in West Bengal on the eastern coast. The eastern coast is generally characterized by a wider coast line when compared to its western counterpart. Coastal ecosystem in India occupies an area of about10.78 million hectares (1,07,833 km^2) extending in nine states. It also occupies considerable area underLakshadweep and Andaman and Nicobar group of Islands. The extent and distribution of coastalarea have been shown in Table 1.

Table 1. Extent and distribution of coastal area in India

State/Union territories	Area (km^2)
West Bengal	14,152
Orissa	7,900
Andhara Pradesh	35,500
Tamilnadu	7,424
Kerala	7,719
Karnataka	7,424
Maharastra	10,000
Goa	220
Gujarat	17465
Lakshadweep	26
Pondicherry and karaikal	3
Total	1,07,833

(Maji *et al.*, 2010)

The coastal tracts of India have a complex geology, with lithological units ranging from recent fluvial and marine deposits to Archaean Crystalline rocks. On the east coast, thickness of alluvium is several hundred metres near the mouths of the major rivers like Cauvery, Krishna, Godavari, Mahanadi, Subarnarekha and Ganga. The western coast, on the contrary has a narrow strip of recent alluvium and older unconsolidated deposits for a major part of its length. The Western Coastal Plains is a thin strip of coastal plain of 50 km in width sandwiched between the Western Ghats and the Arabian Sea. The Eastern Coastal Plains refer to a wide stretch i.e., 100 to 130 km of landmass of India, lying between the Eastern Ghats and the Bay of Bengal.

The coastal area offers a wide variety of coastal and marine resources essential for India's economic growth. Cultural and archaeological sites of prehistoric and/or historic significance dot the eastern and western coasts. Despite the richness and diversity of coastal resources and the capacity to sustain many different forms of economic development, coastal areas are prone to several natural hazards like storms, cyclones, tidal surges, floods, coastal erosion etc. which brings about large scale destruction of life, property and natural resources in the coastal regions of the country every year. The coastal zone represents the transition from terrestrial to marine influences and vice versa. It comprises not only shoreline ecosystems, but also the upland watersheds draining into coastal waters, and the near shore sub-littoral ecosystems influenced by land-based activities. Functionally, it is a broad interface between land and sea that is strongly influenced by both Soils which contain excess soluble salts that affect plant growth adversely are called as salt affected soils. Allsoils contain some of amount of soluble salts. If the soildominated by NaCl and Na_2SO_4, then the soil is said to be saline soil. Salinity range of coastal area (EC) ranges from 0.5dSm-1 in monsoon to 50 dSm^{-1} in summer. If the quantity ofsoluble salts in soil exceeds a particular value, the growth and yield of the crops will be affected adversely, the extent depending on the kind and amount of salts present. The saltreaches the soil surface through capillary rise during dryseason and makes the soil saline and unproductive for agriculture (Prasenjit Ray *et al.*, 2014). Mostly NaCl followed by Na_2SO_4 are the dominant soluble salts, with abundance of soluble cations in the order of Na > Mg > Ca > K.

Coastal Land Salinity- Issues

Salinization is a global issue of serious concern because it reduces the potential productivity and use of land and water resources. In coastal regions, that are in close proximity to the sea, salinization may lead to changes in the chemical composition of natural water resources thus degrading the quality of water supply to the domestic, agriculture and industrial sectors, loss of biodiversity, taxonomic replacement by halo-tolerant species, loss of fertile soil, collapse of agricultural and fishery industries, changes in local climatic conditions, and creating health problems; Thus, affecting many aspects of human life and posing major hindrance to the economic development of the region

The problem of coastal soil salinity in India encompasses

1. Coastal saline soils situated in humid and sub-humid areas
2. Coastal saline soils of arid areas and
3. Coastal acid saline soils.

Salt affected soils occur within a narrow strip of land adjacent to the coast and up to 50 km wide. These areas generally have an elevation of less than 10 m above mean sea level and include the low-lying land of river deltas, lacustrine fringes, lagoons, coastal marshes, and narrow coastal plain or terraces along the creeks. Among the multiple vulnerabilities low-lying coastal areas face due to climate change, progressive water and soil salinization pose particularly serious threats. Worldwide, about 600 million people currently inhabit low elevation coastal zones. Increasing salinity from saltwater intrusion will threaten their livelihoods and public health through its effects on agriculture, aquaculture, infrastructure, coastal ecosystems, and the availability of fresh water for household and commercial use. Therefore, understanding the physical and economic effects of salinity diffusion, and planning for appropriate adaptation, will be critical for long-term development and poverty alleviation in countries with vulnerable coastal regions. Sea-level rise, storm surges, and cyclones exacerbated by climate change have begun to severely affect coasts and river estuaries in low-income countries. The resulting increased salinity in soil and drinking water has health implications for large populations.

Definition of saline soil

The soils which contain sufficient water-soluble salts in the root zone affect or impair plant growth in general, may be termed as saline soil. According to Biswas and Mukherjee (1987), saline soils are soils containing excess of neutral soluble salts dominated by chlorides and sulfates that affect plant growth.

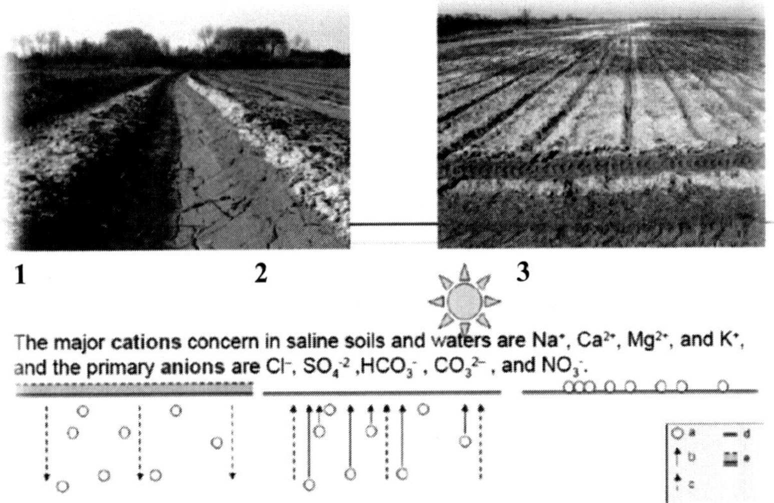

Rain or irrigation in the absence of leaching, can being salts to the surface by capillary action

(Amir Naghverdi *et al.*, 2018)

Field diagnosis of saline soil

Plants may appear water stressed, poor germination, leaf burn, shallow water table and white alkali on the surface of soil are some of the visible symptoms of saline soil.

Physicochemical characteristics of saline soil

EC of soil saturation paste will be > 4 dSm^{-1}, pH of the soil will be < 8.5 and ESP will be < 15. Good soil structure, high rate of infiltration, good soil aeration and white colour of soil are the characteristic features of saline soil.

Major Coastal Soils and Their Formation

Coastal soils are rich in salts, mainly due to the presence of saline ground water table at shallow depth and frequent brackish water inundation in the low lying areas. The ground water influenced by sea and brackish water estuaries reaches the soil surface through capillary rise during dry season, evaporate from the soil leaving salts behind, finally making the soil saline and unproductive for agricultural crops. The soil salinity thus shows high temporal and spatialvariation depending on the elevation, soil texture, climate (evapo-transpiration, precipitation,wind velocity, relative humidity etc.), drainage and other related factors. The salt-laden sands blown by sea winds are greatly responsible for formation of coastal salt-affected soils.

The presence of salinity in soil and water can affect plant growth in three ways

It can increase the osmotic potential and hence decrease water availability. Because of high osmotic pressure, plant roots find difficulty in absorbing high quantity of water and it is due to the presence of soluble salts in soil.The osmotic effect increases the potential forces that hold water in the soil and makes it more difficult for plant roots to extract water. During dry period, salts in soil solution may be so concentrated as to kill plants by pulling water from them (exosmosis). Due to high salt concentration, plants have to spend more energy to absorb water and only a smaller quantity of energy is left for growth thus, seriously affecting cell elongation, making cells flaccid and loss of turgidity of the cell.

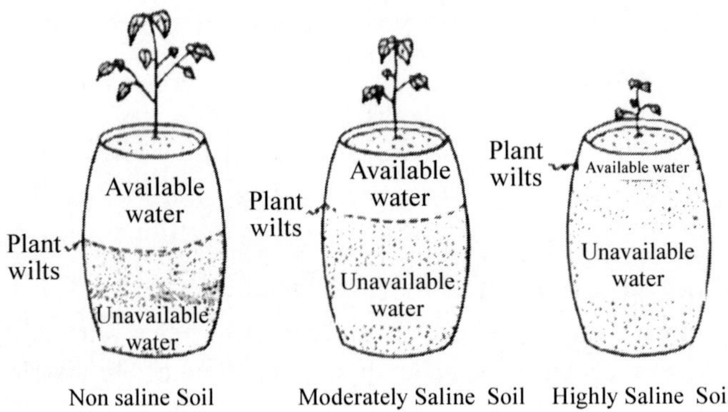

(Groenevelt *et al.*, 2004)

I. Specific-ion effects

1. At low concentration: $NaHCO_3$ and soluble borates become toxic.
2. The harmful effect of $NaHCO_3$ is more likely to be due to the consequences of high pH it brings about. Phophate, Fe, Zn and Mn become unavailable to the plant at high pH value and soil structure tends to become water unstable bringing about conditions of low water permeability and poor aeration.

II. Nutritional Imbalance

1. **HCO_3 - induced Fe deficiency:** Fe is precipitated due to presence of high bicarbonate.
2. **Na+ induced Ca deficiency:** The specific effects of Na on plant physiological processes include antagonistic effects on Ca uptake and shows Ca deficiency. This is because Na^+ displaces Ca^{2+} from membranes, rendering them non-functional.

3. **Mg induced Ca deficiency**

   High concentrations of competitive cations such as Na+, K+ and Mg^{2+} have been shown to displace cell membrane-associated Ca^{2+}. The greater antagonistic effect of Mg^{2+} compared to Na+ is due to its greater membrane binding constant. Due to its greater binding constant, Mg^{2+} more readily displaces Ca^{2+} from the plasma membrane at lower concentrations (and salinities) than Na+, resulting in a greater growth reduction and corresponding Ca deficiency.

III. Sea Water Intrusion

Sea water intrusion is the migration of saltwater into freshwater aquifers under the influence of groundwater development. (Freeze and Cherry, 1979).

Natural Occurrence of Seawater Intrusion

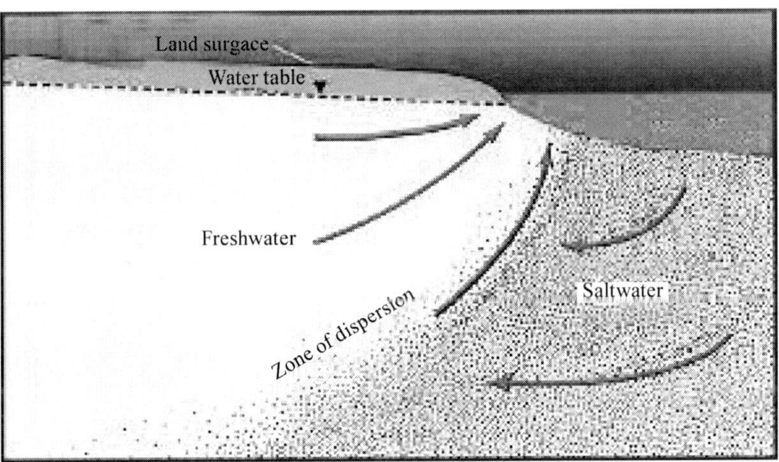

(https://waterusgs.gov/ogw/gwrp/salt water/salt html)

Causes of Sea water intrusion

1. Ground Water Extraction
2. Canals and Drainage networks
3. Higher seawater density than freshwater

1. Ground Water Extraction

- Groundwater extraction is the primary cause of saltwater intrusion.
- It is caused by excessive pumping

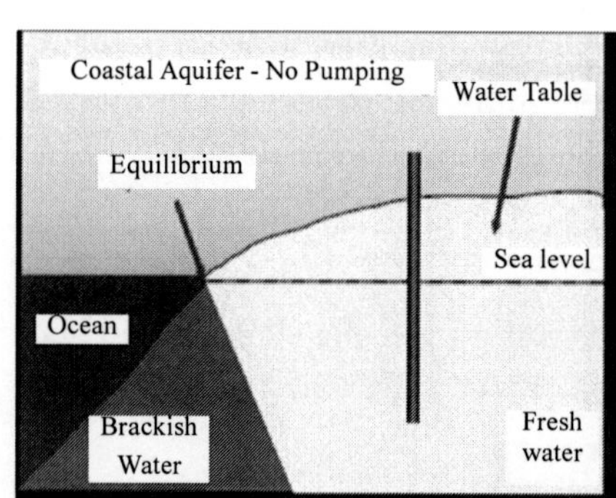

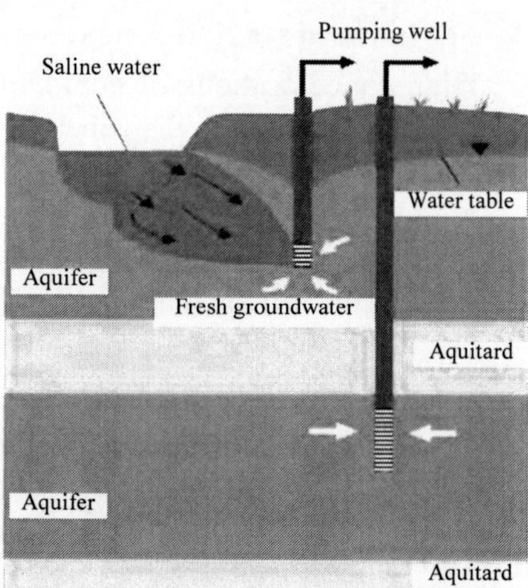

(https://waterusgs.gov/ogw/gwrp/salt water/salt html)

2. Canals and Drainage networks

The construction of canals and drainage networks can lead to salt water intrusion. Canal provide conduits for the salt water to be carried inland as does the deepening of channels for navigation purposes. Drainage networks constructed to drain flat coastal areas can lead to intrusion by lowering freshwater tables, reducing water pressures exerted by fresh water column.

3. High sea water density than fresh water

Fresh water density is 1.0 g/cm^{3,} whereassaltwater density is 1.025 g/cm^3. The law of nature is that objects with higher density flows towards the object with lower density. Thus the sea water which has higher density than the fresh water will flow towards the fresh water area when the water from there is removed because of extraction.

Initiatives to mitigate coastal salinity

I. Preventive Measures

The preventive measures include establishing salinity coastal techniques like

1. Tidal regulators and Bandharas at river
2. Physical barriers
3. Freeh water barriers
4. Extraction barriers.

II. Management approaches to increase productivity of coastal soils

1. **Leaching the soil**: The salinity level of salt-affected coastal soils can be reduced by leaching the soils with good quality water. This can be a better option to reclaim the cyclone affected soils of the coastal area also.

In the low-lying coastal areas where water table remains shallow for most part of the year and the quality of ground water is poor, installation of sub-soil drainage system is more useful.

2. **Soil Management:** Soil management techniques generally recommended include (a) Maintenance of satisfactory fertility levels, pH and structure of soils to encourage growth of high yielding crops; (b) Maximization of soil surface cover, e.g. use of multiple crop species; (c) Mulching to help retain soil moisture and reduce erosion; (d) Crop selection, e.g. use of deep-rooted plants to maximise water extraction; (e). By using crop rotation, minimum tillage and minimum fallow periods.
3. **Water Management:** The following water management practices can be followed to mitigate coastal soil salinity. (a) Efficient irrigation of crops, soil moisture monitoring and accurate determination of water requirements; (b) Choice of appropriate drainage methods according to the situation: (*i*). Surface drainage systems to collect and control water entering and/or leaving the irrigation site; (*ii*). Subsurface drainage systems to control a shallow water table below the crop root zone; (*iii*). Bio drainage: the use of vegetation to control water fluxes in the landscape through evapo-transpiration and (*iv*). Adequate disposal of drainage water to avoid contamination of receiving waters and the environment.
4. **Appropriate use of ridges or beds for planting:** The impact of salinity may be minimized by placing the seeds (or plants) appropriately on ridges. The place where exactly the seeds should be planted on the ridge or bed will depend on the irrigation design. If the crop planted on ridges would be irrigated via furrows on both sides of the ridge, it is better to place plants on the ridge shoulders rather than the ridge top because water evaporation will concentrate more salts on the ridge top or center of the bed.

   Sloping beds may be slightly better on highly saline soils because seeds can be planted on the slope below the zone of salt accumulation.

   Pattern of salt build-up as a function of seed placement, bed shape and irrigation water quality.(Bandyopandhaya *et al*., 1988)

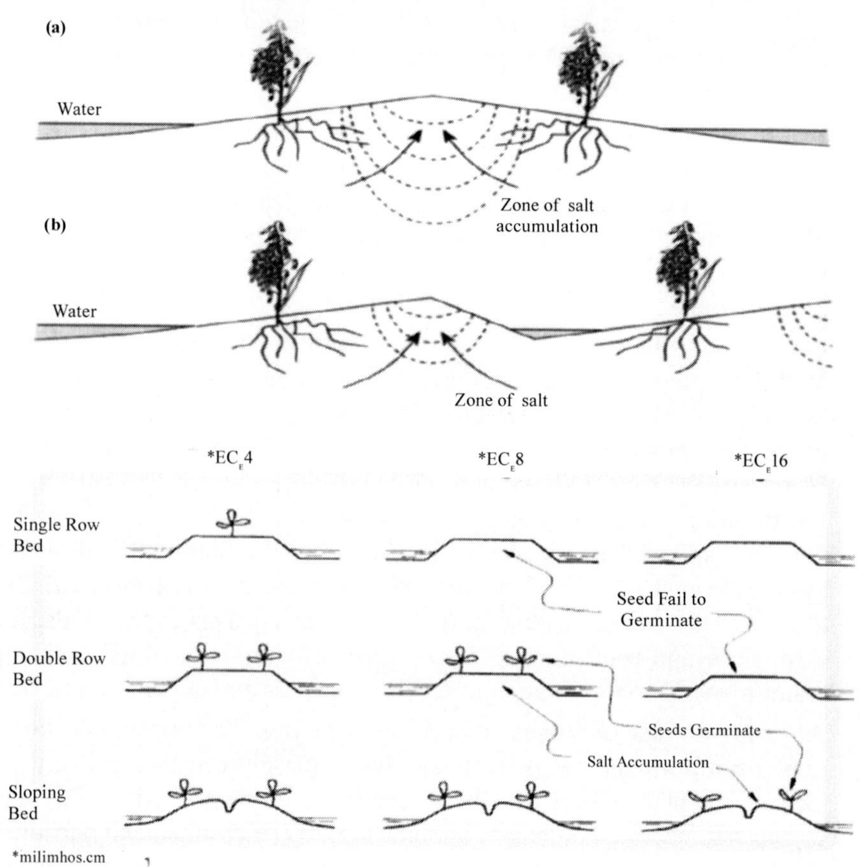

5. Mulching

Mulching with crop residue, such as straw, reduces evaporation from the soil surface which in turn reduces the upward movement of salts. Reduced evaporation also reduces the need to irrigate and consequently only fewer salts accumulate.

6. Deep Tillage

Accumulation of salts closer to the surface is a typical feature of saline soils. Deep tillage would mix the salts present in the surface zone into a much larger volume of soil and hence reduce its concentration and impact. Many soils have an impervious hard pan which hinders in the salt leaching process. Under such circumstances "chiseling" would improve water infiltration and hence there will be downward movement of salts into the deeper layers of soil.

7. Incorporation of Organic matter

Incorporating crop residues or green-manure crops improves soil tilth, structure, and improves water infiltration which provides safeguard against adverse effects of salinity. In order for this to be effective, regular additions of organic matter (crop residue, manure, sludge, compost) must be made.

8. **Paddy-cum-fish culture**

For effective utilization of land resources and betterment of livelihood of the local people in coastal areas, paddy-cum-fish culture can be adopted in lowland areas of coastal regions without much affecting the productivity of the soils. This may facilitate additional income generation to the farmers struggling for survival in the coastal regions and also uplift their socio-economic condition.

Land shaping for broad bed furrow system of cultivation
(Sabareshwari and Ramya, 2018)

Conclusion

Coastal areas are different from inland areas. Salinity problems in coastal soil is caused during the process of their formation under marine influence and subsequently due to periodical inundation with tidal water and in case of lowlands having proximity to the sea, due to the high water-table with high concentration of salts in it. The coastal soils exhibit a great deal of diversity in terms of climate, physiographic and physical characteristics, as well as in terms of generally a rich stock of flora and fauna. In coastal and delta regions of major rivers of the world, the soils are rich in salts due to the presence of saline ground water-table at shallow depth.

The coastal delta regions are usually having low elevation and hence may also be subjected to frequent brackish water inundation. The ground water in the region is influenced by sea, the brackish water estuaries. The soil salinity thus shows high temporal and spatial variation depending on the elevation, soil texture, climate (evapo-transpiration, precipitation, wind velocity, relative humidity, etc.) drainage and other related factors. The ingress of sea water or brackish water, and salt-laden sands blown by sea winds are also greatly responsible for formation of coastal salt affected soils. Improved technologies for higher productivity is achieved through drainage embankments, watershed management, rainwater conservation, improved soil amelioration and other management practices such as leaching of salts, shifting cultivation and improved fertilizer management.

References

Amir Haghaverdi, W. Laosheng,V., Janet Hartin and Laurent Ahiablame. (2018). Soil salinity. Digital Edition, North Star Publishing Inc, New York.

Bandyopandhaya, A.K., Bhargava, G.P., G, K.V., Rao, K., Sen,H.S., Sinha, T. S., Bandopadhyay, B. K., Biswas,C.K., Bal, B.R., Dutt, S.K. and Mondal, R.C. (1988). Coastal saline soils of India and their management. CSSRI, Karnal.

Biswas, T. D. and Mukherjee, S. K. (1987). Textbook of Soil Science. Tata McGraw Hill Publishing Company, New Delhi.

Freeze, R.A. and Cherry, I.A.(1979). Ground water. Prentice Hall, Inc, New York.

Groenevelt, P. H., Grant, C.D., and Murray, R.S. (2004). Water availability in saline soils. *Aust. J. Soil Res.*, **42(2):** 833–840.

Maji, A.K., Obi Reddy, G.P. and Dipak Sarkar. (2010). Degraded and wastelands of India. Status and spatial distribution, ICAR, New Delhi.

Prasenjit Ray, Babu Lal Meena and Nath, C. P. (2014). Management of Coastal Soils for Improving Soil Quality and Productivity. *Popular Kheti,* **2(1):** 95–99.

Sabareshwari, V. and Ramya, A.(2018). Coastal saline Soils of India: A review. *Agricultural Reviews*, **39(1)**: 86–88.

9

Impact of Climate Change on Agricultural Production

S. Panccrsclvam
Professor and Head, Agro Climate Research Centre
Tamil Nadu Agricultural University, Coimbatore

Agriculture is dependent directly on weather and indirectly on the consequences of climate change. Changes in temperature, precipitation, incoming solar radiation and CO_2 concentration are expected to impact crop growth to a greater extent. The overall impact of climate change on worldwide food production is considered to be low to moderate only when there is successful adaptation strategies towards climate change (IPCC, 1998). Global agricultural production could be benefited by the doubling CO_2 concentration. All sectors including agriculture will be impacted by the adverse effect of climate change imposed on water resources (Gautam and Kumar, 2007; Gautam, 2009). India is expected to face more seasonal variation in temperature with more warming in the winters than summers (Christensen *et al.,* 2007; Cruz *et al.,* 2007). The frequency of drought occurrence is increasing due to variations in the monsoonal rainfall. Climate change is posing a great threat to agriculture and food security. Water is the most critical agricultural input in India, as 55% of the total cultivated areas do not have irrigation facilities.

Even with the variations in climate conditions, food accessibility for the present generation is assured, but as far as the future is concerned there is still the uncertainity of agricultural production. All climate models predict that there will be more extreme weather conditions, with more droughts, heavy rainfall and storms in agricultural production regions. Such extreme weather events will influence pest and disease occurance, imposing severe risks of crop failure. In India, climate change could be a potential threat towards food security due to population growth, industrialization and economic development. India's climate could become warmer under conditions of increased atmospheric carbon dioxide.

Factors affecting crop productivity

Rainfall is the major deciding factor of agriculture production. A warmer climate will accelerate the hydrologic cycle, altering rainfall, magnitude and timing of runoff. Warming of air temperature makes it possible to increase the evapotranspiration (ET) demand. In arid regions of Rajasthan state, an increase of 14.8 per cent in total ET demand has been projected with increase in temperature (Goyal, 2014). In addition, rise in sea level will increase the risk of permanent or seasonal saline intrusion into ground water and rivers which will have an impact on quality of water and its potential use of domestic, agricultural and industrial uses. Climate change will have number of effects on agriculture (Gautam and Sharma, 2012).

Higher temperatures will severely affect the production patterns of different crops. The cropping zones may g*et al*tered due to the increasing temperature and variation in rainfall. Wheat yields are predicted to fall by 5–10% with every increase of 1°C and overall crop yields could decrease up to 30% in South Asia by the mid-21st century (IPCC, 2001). India could experience a 40% decline in agricultural productivity by the 2080s (IPCC, 2007). Higher growing season temperatures can significantly impact agricultural productivity, farm incomes and food security (Battisti and Naylor, 2009). In mid and high latitudes, the suitability and productivity of crops are projected to increase and extend northwards, especially for cereals and cool season seed crops. An increase in the mean seasonal temperature can bring forward the harvest time of current varieties of many crops and hence reduce final yield without adaptation to a longer growing season. In areas where temperatures are already close to the physiological maxima for crops, such as seasonally arid and tropical regions, higher temperatures may be more immediately detrimental, increasing the heat stress on crops and water loss by evaporation.

Radiation frosts are common in many temperate environments due to high radiative loss on cloudless, clam nights. Intracellular freezing ensure at around –7°C, tissue death resulting from the combined effects of membrane injury, cytoplasm dehydration and protein denaturation. Frost hardiness of plant cells involves cell size, wall thickness, osmotic pressure of cell sap and membrane properties, all of which can either delay the onset or diminish adverse consequences of ice formation. Plant organs that are rapidly growing rapidly are sensitive to frosts.

Theoretical estimates suggest that increasing atmospheric CO_2 concentrations to 550 ppm, could increase photosynthesis in such C_3 crops by nearly 40 per cent (Long *et al.*, 2004). The physiology of C_4crops, such as maize, millet, sorghum and sugarcane is different. In these plants CO_2 is concentrated to three to six times atmospheric concentrations and thus

RuBisCO is already saturated. Thus, rising CO_2 concentrations confer no additional physiological benefits. These crops may, however, become more water-use efficient at elevated CO_2 concentrations as stomata do not need to stay open as long for the plant to receive the required CO_2. Thus yields may increase marginally as a result (Long *et al.*, 2004).

Adaptation strategies

- **Developing tolerant varieties**- Drought and Salinity tolerant varieties of crops are the need of future due to the lack of required quantity and quality of irrigation water.
- **Crop diversification**- Relying on a monocrop can worsen the situation of crop failure. Crops that are more suited to adversities of climate change are present benefitters to the farming community.
- **Change in cropping pattern and planting dates**- Altering the cropping pattern and date of planting according to the monsoonal forecast can help farmers to coincide with the rainfall days. Contingent Crop Planning by the State Government plans towards making benefit of the available rainfall to the maximum use. Mixed type of farming can help in coping up with the loss on overall sector.
- **Improving irrigation methods**- With the scarcity of water becoming a potential threat to agriculture, the efficiency has to be increased from usage side. More focus has to be given towards drip and micro irrigation methods where water directly is used by the plant without any wastage. Flood irrigation in case of rice could no longer be relied on the rainfall or water from rivers. Methods like alternate wetting and drying are to be taken into serious consideration to save the existing water resources and prevent crop failure.
- **Soil moisture conservation**- The available soil management practises should focus on soil moisture conservation aspects. Usage of mulches, zero tillage, soil amendments etc can help in conserving soil moisture.
- **Agroforestry**- More stable and reliable income could be generated with agroforestry management. Planting tress turn out to be environmentally friendly and cultivating the available land will generate the additional income to sustain farm income. This turn out to be win-win situation for both the farmer and the ecology.

Mitigation strategies

- Improved crop and grazing land management to increase soil carbon storage.
- Restoration of cultivated peaty soils and degraded lands.

- Improved rice cultivation techniques and livestock and manure management to reduce CH_4 emissions.
- Improved nitrogen fertilizer application techniques to reduce N_2O emissions.
- Dedicated energy crops to replace fossil fuel use.
- Improved energy efficiency.

References

Battisti, DS. and Naylor, RL. (2009). Historical warnings of future food insecurity with unprecedented seasonal heat. *Science,* **323 (5911):** 240–244.

Christensen, J.H., Hewitson, B., Busuioc, A., Chen, A., Gao X. (2007). Regional Climate Projections. In: Climate Change (2007): The Physical Science Basis. Cambridge University Press. Cambridge, United Kingdom.

Cruz, RO., Harasawa, H., Lal, M., Wu, S., Anokhin, Y. *et al.* (2007). Asia Climate Change.(2007): Impacts, Adaptation and Vulnerability. Contribution of Working Group II to the Fourth Assessment Report of the Intergovernmental Panel on Climate Change. Cambridge University Press, Cambridge, UK.

Gautam, H R. (2009). Preserving the future. In; Joy of Life- The Mighty Aqua". Bennett, Coleman & Co. Ltd., the Times of India, Chandigarh.

Gautam, HR.and Kumar, R. (2007). Need for rainwater harvesting in agriculture. *J Kurukshetra*, **55**: 12–15.

Gautam, HR. and Sharma, HL. (2012). Environmental degradation, climate change and effect on agriculture. *J Kurukshetra*, **60**: 3–5.

Goyal, RK. (2004). Sensitivity of evapotranspiration to global warming: a case study of arid zone of Rajasthan (India). *Agric Water Manage*. **69**: 1–11.

IPCC. (1998). Principles governing IPCC work, Approved at the 14th session of the IPCC.

IPCC. (2001). Climate Change: Impacts, Adaptation & Vulnerability: Contribution of Working Group II to the Third Assessment Report of the IPCC. Cambridge University Press, Cambridge, UK.

IPCC. (2007). Summary for Policy-makers, Climate Change: Mitigation. Contribution of Working Group III to the Fourth Assessment Report of the IPCC. Cambridge University Press, Cambridge, United.

Long, S. P., Ainsworth, E. A., Rogers, A. and Ort, D. R. (2004). Risingatmospheric carbon dioxide: Plants face the future. *Ann. Rev. Plant Biol.*, **55(1)**: 591–628.

10

Impact of Climate Change in Horticultural Crops

S. Ramesh kumar, D. Dhanasekaran and R. Jeya
Faculty of Agriculture, Annamalai University, Annamalai Nagar, Tamil Nadu 608002

Global warming has become a worldwide concern after feeling the rise of temperature much more clearly after 1990. Global warming has negatively affected agricultural production throughout the world attributed by the climate change parameters. Effects of climate change on horticultural crops reported nowadays include problems of biotic and abiotic stress, yield loss, quality loss, etc. To understand the impact of climate change problems in horticulture, it is prerequisite to comprehend the issues related to global carbon cycle, climate change phenomenon, relationship between climate change variables and direct as well as indirect climate change effects.

Global Carbon Cycle

The near-surface environment of the Earth contains approximately 121,000,000 gigatons of carbon (GtC). Based on its availability in the atmosphere, the carbon is divided into three types. They are,

1. Carbon that is locked away in permanent storage (It is not available to combine with oxygen and form CO_2 in the atmosphere).
2. Carbon that is in relatively long-term storage in the land and the ocean and,
3. Carbon that is already in the atmosphere (mainly as CO_2 gas).

About 78,000,000 GtC or two-thirds of the near-surface carbon on Earth occurs in nearly permanent storage in fossil fuels, limestone rocks, or sediments. Most of this carbon was originally in the atmosphere but has gone into storage underground over millions of years. Most of the remaining one-third (44,000 GtC or one-third of the total) is in relatively long-term storage in the ocean and at the surface of the land. A small part of the carbon, only 750 GtC (less than 1% of all the near-surface carbon on the Earth) occurs in the form of a gas in the atmosphere. Most of this carbon is combined with oxygen as the gas CO_2.

Each year, about 260 GtC (0.35% of the carbon in relatively long-term storage) moves from the land and ocean to the atmosphere, and a nearly equal amount moves from the atmosphere into temporary storage in the ocean and the land. This cycle has been relatively constant, but there have been times in the past when CO_2 levels in the atmosphere have been relatively high. There have also been periods when the amount of CO_2 in the atmosphere has been relatively low (UNDEERC, 2018). When, the amount of carbon present in near surface environment is plotted in a graph along with the amount of carbon involved in carbon cycle, the later data becomes invisible, yet it has greater influence in global warming process (Fig. 1).

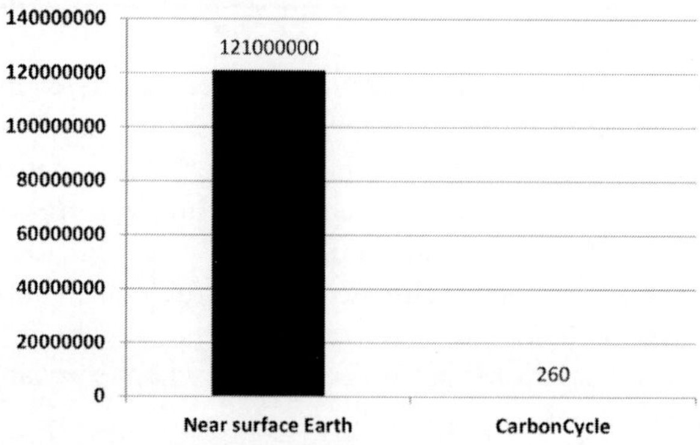

Figure 1. Amount of Carbon in Near Surface Earth VS Carbon Cycle (in GtC)

Climate Change

The climate on the Earth depends on a number of factors, including the mix of gases in the atmosphere, the amount of energy the Earth receives from the sun, and the conditions at the Earth's surface. Scientists have shown that the levels of carbon dioxide (CO_2), a major greenhouse gas (GHG) in the atmosphere, have significantly changed over the last 600 million years and that temperature has varied as well. Since the early 1800s, atmospheric concentrations of CO_2 have increased nearly 30 per cent and the concentrations of other GHGs like methane (more than doubled) and nitrous oxide (up by about 15%) have also increased (EIA, 2004). The CO_2 in the atmosphere at present are higher than they have been at any time in the past 400,000 years. During ice ages, CO_2 levels were around 200 parts per million (ppm), and during the warmer interglacial periods, they hovered around 280 ppm. In 2013, CO_2 levels surpassed 400 ppm for the first time in recorded history (NASA, 2018). It is agreed fact that CO_2 levels in the atmosphere are currently high and are increasing. The average annual temperature at the surface of the Earth has increased by about 1°F over the past 150 years.

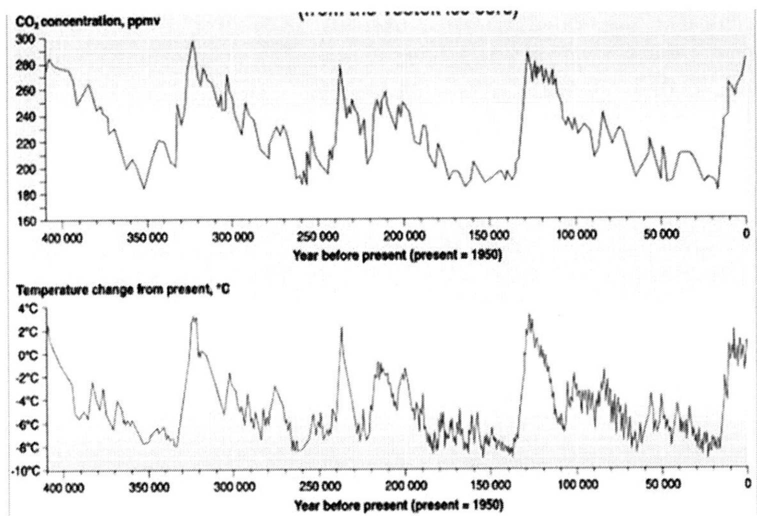

Figure 2. Temperature and CO_2 concentration in the atmosphere over the past 400,000 years (From the Vostok ice core, *Source*: Petit *et al.*,1999)

Petit *et al.* (1999) after observing changes in atmospheric CO_2 and Antarctic temperature suggested that the oceanic area around Antarctica plays a role in the long-term CO_2 change. CO_2 and CH_4 concentrations are strongly correlated with Antarctic temperatures. Their results support the idea that greenhouse gases have contributed significantly to the glacial–interglacial change. This correlation, together with the uniquely elevated concentrations of these gases today, is of relevance with respect to the continuing debate on the future of Earth's climate.

Relationships between major climate variables

As climate constitutes a complex of variables, their relationship need to be understood clearly. The term climate change refers to long-term changes in temperature, humidity, clouds and rainfall and not to day-to-day variations (IPCC, 2007). A smooth change in one variable triggers smooth changes in most others. The CO_2 and other greenhouse gases play a part largely through their effect on the radiation balance of the atmosphere. There is only a weak link between such factors as cloudiness and wind. Temperature, evaporation and rain are strongly correlated, which illustrates the likely intensification of the hydrological cycle (Figure 3). Combined with the projected pressure on land and water use, competition for land and water will certainly become a key social and political issue. Climate variability is likely to increase under global warming (Katz and Brown, 1992), both in absolute and in relative terms. This is linked with thresholds which affect the occurrence of many meteorological phenomena. For instance, tropical cyclones are 'fed' by water vapour evaporating from oceans at a temperature above 26 or 27°C. Therefore, higher average sea surface temperatures are bound to result in a higher frequency of tropical cyclones (Wim and Rene, 1996).

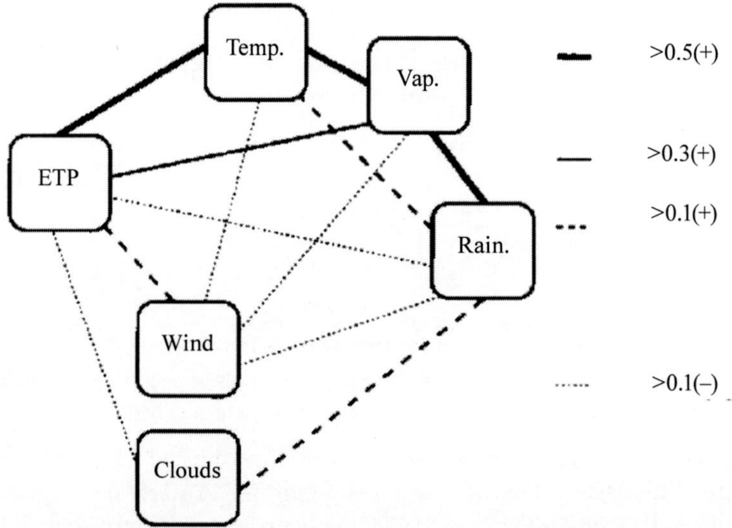

Figure 3. Relationships between major climate variables (Source: FAO, 1995) (Temperature-*Temp.,* Vapour pressure, -*Vap.,* Rainfall –*Rain,* Wind speed –*Wind,* Cloudiness-*Clouds* and Evapo-transpiration potential-*ETP*)

Change in Frequency of Cyclones

The climate hazard caused by a tropical cyclone is not just due to winds, but also due to both coastal and freshwater flooding caused by heavy rain linked to tropical cyclones. The severity of cyclones in Bay of Bengal has triggered speculations about whether it is linked to climate change. But the impact of climate change on intensity and frequency of cyclones is not understood well and so far not been proved conclusively. It has been observed that there is an increase in numbers and proportion of hurricanes of the category 4 and 5 globally since 1970 with a simultaneous decrease in the total number of cyclones and cyclone days. Some climate models predict two to seven fold increase in the frequency of Hurricane Katrina magnitude events for a 1°C rise in global temperature. Climate models mostly continue to predict future decreases in global tropical cyclone numbers, projected increases in the intensities of the strongest storms and increased rainfall rates. In contrary, Pielke and Land sea (1998) opined that tropical cyclones have become increasingly frequent and severe over the last four decades as climatic conditions have changed in the Tropics of United States.

Akhila and Annadurai (2018) reported that in India, West Bengal, Orissa and Andhra Pradesh, Andaman and Nicobar Islands are mostly affected states by the tropical cyclones from the plots. Gujarat and Lakshadweep are mostly affected with storms that are originating in the Arabian Sea. Tropical cyclones suddenly change their course of path during their active life span which is also observed while plotting the track of cyclones. The main cyclone season in the South Indian Ocean observed is May-July and September-December with major occurrences of storms in April and August from the historical data.

Impact in low-lying coastal zones

The low-lying coastal zones are more vulnerable to submergence or salt water intrusion caused by rising sea levels that ultimately affects the agriculture activities. The estimated 40cm sea-level rise in the coming 100 years would submerge some valuable coastal agricultural lands which threatens the future of coastal horticulture. The physical changes expected due to sea level rise are,

- Submergence of low-lying cultivable land areas
- Erosion of soft shores by increasing offshore loss of sediment
- Increased salinity of estuaries, irrigation channels and aquifers
- Rising coastal water tables which can affect irrigation sources and
- Severe coastal flooding and storm damage.
- Soil conditions may pose problems with an increase in acidity, alkalinity and salinity are expected. Coastal regions can expect much faster percolation of sea water in inland water tables causing more salinity.

Biodiversity loss

Global climate change is often considered as one of the major factors causing biodiversity loss. A large portion of biodiversity including plant species of horticultural importance is abundant in mountains. Mountain ecosystems are often endemic, because many species remain isolated at high elevations compared to lowland vegetation communities that can occupy climatic niches spread over wider latitudinal belts (Beniston, 2003). Their geographical isolation, limited range size and unique environmental adaptations make montanne species potentially the most threatened under impeding climate change (La Sorte and Jetz, 2010). As endemic species often show narrow altitudinal distribution patterns, they are more threatened (Grabherr *et al.*, 1994). The horticultural biodiversity in high-altitude ecosystems are largely controlled by climatic constraints, and many plants occur close to their climatic limits of survival. Changes in the habitats where species occupy and changes to the composition of plant and animal communities are considered as key drivers of climate change on biodiversity in a place. For example, the highly bio diverse but fragile mountain ecosystems of North East India have diverse vegetation types encompassing from the subtropical, submontanne, montanne, subalpine to alpine systems. North East India is nestled in the globally recognized biodiversity hotspot renowned for its high species diversity and endemism. It is also recognized as one of the centres of origin of cultivated plants. Change in temperature regime may cause severe impact and significant changes in rich horticultural biodiversity of North East India. Loss of biodiversity from some of the most fragile environments, such as tropical forests and mangroves is also most common.

Impact in pest and disease management

The incidence of diseases and pests, especially alien ones, could increase. The elevated CO_2 levels can cause increase in simple sugars and reduction in nitrogen content of the leaves. This condition in leaves may lead to more damage caused by many insects which consume more leaves to meet their metabolic requirements of nitrogen. Thus, any insect attack will be more severe. Higher temperatures will lead to a poleward spread of many pests and diseases in both hemispheres. This will lead to more attacks over longer periods in the temperate climatic zone (Venkataraman, 2006). Change in the pest complex that attack crop plants due to climate change has also been envisioned. Reports state that some indigenous pests that were earlier not causing much damage are emerging as serious pests such as bract mosaic in banana, etc. Similarly, the advantage of an aphid-free-season will be lost in the case of potato seed production in North India. Models on plant diseases indicate that climate change could alter stages and rates of development of certain pathogens, modify host resistance, and result in changes in the physiology of host-pathogen interactions. The most likely consequences are shifts in the geographical distribution of pathogens and increased crop losses (Swaminathan and Kesavan, 2012). For example, Fruit flies can cause extensive damage to fruits and vegetables production. Controlling such pests often require the use of pesticides, which can have serious side effects on human health and the environment due to unsafe levels of pesticide residue on food supply. Increased temperature could decrease insect population (eg. aphids) in some crops, which cannot be grown in higher temperature. The same condition may be conducive for increased activity of natural enemies of that pest, thereby further reducing its population. The high CO_2 level in atmosphere has indirect impact on insect population. In certain crops, higher CO_2 concentration had 57% more insect damage (Japanese beetle,

Leafhopper, Root worm, Mexican bean beetle) than earlier. It causes increase in level of simple sugars in the leaves that stimulates more feeding by insects. Increased C/N ratio in plant tissue due to increased CO_2 level may slow insect development and increase life stages of insect pests vulnerable to attack by parasitoids.

Direct Plant physiological effects of elevated CO_2

The greenhouse gases CH_4, N_2O and chlorofluorocarbons (CFCs) have no known direct effects on plant physiological processes. However, they influence change in global temperature. Elevated CO_2 has direct influence on plant physiological processes and also indirectly involved through increased UV-B through depleted stratospheric ozone, increased temperatures and the associated intensification of the hydrological cycle.

The CO_2 fertilization effect

CO_2 is an essential plant 'nutrient', in addition to light, suitable temperature, water and chemical elements such as N, P and K, and it is currently in short supply. Higher concentrations of atmospheric CO_2 due to increased use of fossil fuels, deforestation and biomass burning, can have a positive influence on photosynthesis (Figure 4); under optimal growing conditions of light, temperature, nutrient and moisture supply, biomass production can increase, especially of plants with C_3 photo-synthetic metabolism, above and even more below ground.

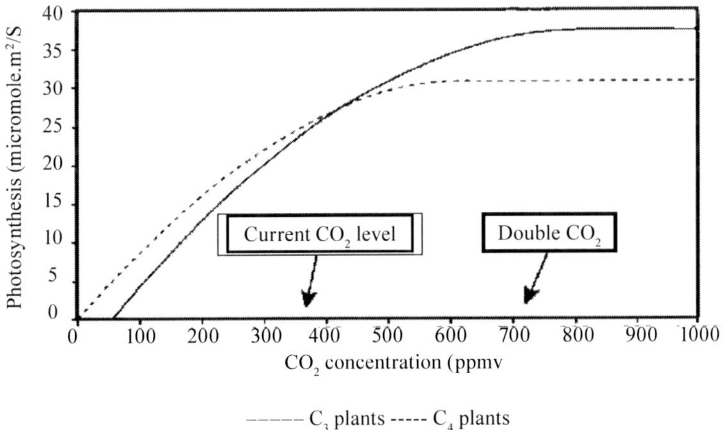

Figure 4. Effect of CO_2 concentrations on C_3 and C_4 plants (Wolfe and Erickson, 1993).

The main mechanism of CO_2 fertilization is that it depresses photo-respiration, more so in C_3 than in C_4 plants.

C_3 species (*viz.,cotton, rice, wheat, barley, soybeans, sunflower, potatoes, most leguminous and woody plants, most horticultural crops and many weeds*) respond more to increased CO_2; C_4 species (*viz., maize, sorghum, sugar cane, millets, halophytes* (*i.e., salt-tolerant plants*) *and many tall tropical grasses, pasture, forage and weed species*) respond better than C_3 plants to higher temperature and their water-use efficiency increases more than for C_3 plants. There are some indications that enhancements can decline over time.

A total of 10 to 20% of the approximate doubling of crop productivity over the past 100 years could be due to this effect (Tans *et al.,* 1990) and forest growth or regrowth may have been stimulated as well. Further productivity increases may occur in the coming century, in the order of 30% or more where plant nutrients and moisture are adequate. Higher CO_2 values would also mitigate the plant growth damage caused by pollutants such as NO_x and SO_2 because of smaller stomatal openings. Higher percentages of starch in grasses improves their feeding quality, implying less need for feed mixes when silaging.

The CO_2 anti-transparent effect (improved water-use efficiency)

With increased atmospheric CO_2, the consumptive use of water becomes more efficient because of reduced transpiration. This is induced by a contraction of plant stomata and/or a decrease in the number of stomata per unit leaf area. This restricts the escape of water vapour from the leaf more than it restricts photosynthesis (Wolfe and Erickson, 1993).

With the same amount of available water, there could be more leaf area and biomass production by crops and natural vegetation. Plants could survive in areas hitherto too dry for their growth. A change in temperature and moisture levels may lead to a change in the absorption rate of fertilizers and other minerals, which determine yield output. In short, the rise in temperature along with the reduction in rainfall reduces agricultural productivity if both are beyond the threshold that is suitable for crop production (Tirado and Cotter, 2010).

Ultraviolet Radiation

Increased ultraviolet radiation (UV-B, between 280 and 320 nanometres), due to depletion of the stratospheric ozone layer, mainly in the Antarctic region, may negatively affect terrestrial and aquatic photosynthesis and animal health. Over the last decade, a decrease of stratospheric ozone was observed at all latitudes (about 10% in winter, 0% during summer and intermediate values during spring and autumn). However, the 'Biological Action Factor' of UV-B can vary over several orders of magnitude with even slight changes in the amount and wavelength of

UV-B (Runeckles and Krupa, 1994). It has been reported that UV-B affects the ability of plankton organisms to control their vertical movements and to adjust to light levels. Reductions in yield of up to 10% have been reported in plants where the CO_2 fertilization effect is strongest. On the other hand, UV-B increase could increase the amount of plant internal compounds that act against pests. The changes in CO_2 tropospheric ozone and increased UV-B do not necessarily occur simultaneously: CO_2 increase is worldwide, but with a strong seasonality in middle and higher latitudes; significant increase of UV-B is largely limited to subpolar regions; high near-surface O_3 levels are restricted to the neighbourhood of major cities, airports, etc. (Seitz, 1994).

The hydrological cycle and soils

Even a slight increase in surface temperatures will affect evaporation, atmospheric moisture and precipitation (Figure 3). While it is generally agreed that rainfall will increase (by an estimated 10 to 15%), two aspects have to be elucidated: how will rainfall intensities be affected, and what are the details of spatial changes. Based on palaeoclimatic analogies, certain authors predict more favourable rainfall conditions in the present-day (Petit-Maire, 1992). If the increase in precipitation should be associated with increased rainfall intensities, then the quality and quantity of soil and water resources would decline, for instance through increased runoff and erosion, increased land degradation processes, and a higher frequency of floods and possibly droughts. The extra precipitation on land, if indeed including present sub-humid to semi-arid areas, will increase plant growth in these areas, leading to an improved protection of the land surface and increased rainfed agricultural production; in already humid areas the extra rainfall may, however, impair adequate crop drying and storage (Wim and René, 1996). The greatest risks are often estimated to be associated with increased soil loss through erosion.

Soils, as a medium for plant growth, would be affected in several other ways (Wim and René, 1996):

- increased temperatures may lead to more decomposition of soil organic matter;
- increased plant growth due to the CO_2 fertilization effect may cause other plant nutrients such as N and P to become in short supply; however, CO_2increase would stimulate mycorrhizal activity (making soil phosphorus more easily available), and also biological nitrogen fixation (whether or not symbiotic). Through increased root growth there would be extra weathering of the substratum, hence a fresh supply of potassium and micronutrients;
- the CO_2 fertilization effect would produce more litter of higher C/N ratio, hence more organic matter for incorporation into the soil as humus; litter

with high C/N decomposes slowly and this can act as a negative feedback on nutrient availability;

- the 'CO_2 anti-transpirant' effect would stimulate plant growth in dryland areas, and more soil protection against erosion and lower topsoil temperatures, leading to an 'anti-desertification effect'.

Effect of Climate Change in Horticulture

Most common indirect effects of climate change on agriculture in general and horticulture in particular caused many negative effects that brought changes in crop production system, productivity and production. However, horticultural crops have certain advantages of mitigating climate change and offer climate resilience.

Increasing CO_2, temperature, humidity and other greenhouse gases might be beneficial in regions where crop production is limited by cold temperatures due to lower potential for frost damage, faster growth and lengthened growing seasons.

Further, there is some benefit coming from CO_2 fertilization, which can enhance photosynthesis. However, climate change could depressingly impact horticulture in regions where climate conditions are currently good for production. For example, climate change might decrease chilling and inhibit bloom and fruit set in horticultural crops, lead to high temperature and wind during bloom or ripening that could negatively impact fruit set or fruit quality, increase evapotranspiration rates that could lead to water deficits, and increased problems with heat stress.

A rise in 1°C temperature may bring in latitudinal and longitudinal shift in the cropping pattern of horticultural crops by influencing changes in major area of potentially suitable zones. Studies conducted at Indian Institute of Spices Research (IISR), Calicut using GIS models have shown that many suitable areas of spices will become marginally suitable or new areas, which are presently unsuitable, become highly suitable for cultivation of spices. This holds good to a variety of horticultural crops.

Because of rise in temperature, crops will develop more rapidly and mature earlier leading to changes in production timing. For example, Citrus, grapes, melons etc., will mature earlier by about 15 days and their availability during late season is restricted. Due to rise in temperature, photoperiods may not show much variation. Hence, photosensitive crops like onions will mature faster leading to small bulb size. In crops like Strawberries, more number of runner shoots are formed instead of fruiting shots due to rise in temperature. The winter regime and chilling duration will reduce in temperate regions affecting the temperate crops.

The faster maturity and higher temperature induced ripening will make the produce, a less storage period in trees/plants. In crops like tomato, pollination will be affected adversely because of stigma drying higher temperature. High temperature may also lead to floral abortions. The requirement of annual irrigation will increase, not only due to higher evaporation but also due to faster biomass production. Heat unit requirement will be achieved in much lesser time leading to shorter duration of annual crops and early flowering in perennial crops.

Stress related problems in horticultural crops due to climate change

Crop	**Production problem**
Potato	High temperature reduced tuber initiation.
Tomatoes	High temperature causes poor pollination, reduced quality of fruits, tip burn and blossom end rot.
Capsicum & apples	High temperature causes problem in anthocyanin production.
Pome and stone fruits	High temperature causes early dormancy breaking.
Banana	Leaf production increases by one leaf per month for every 3.3 to 3.7°C rise in minimum or mean temperature from 10–20°C or 13.5 to 25°C respectively. Bright sunshine causes sunburn damage on exposed fruit. The temperature below 10°C leads to impedance of inflorescence and malformations of bunches. Accumulation of sodium ions due to salt stress causes scorching of leaf margins.
Cassava and sweet potato	Significant reduction in tuber yield as well as in starch content due to drought. In cassava, high CO_2 (700 ppm) and high temperature increased the tuber yield.
Plantation Crops	In coconut, arecanut and cocoa, increased CO_2 led to higher biomass production.
Cashew	Due to unseasonal rains, the nuts and apples at ripening stage are spoiled due to blackening of nuts as well as rotting of apples on trees itself. Flowers exposed to heavy rain accompanied with wind results in flower drop and occurrence of fungal diseases.
SPICES	Shift in the areas of cultivation due to rise in temperature. These shifts would lead to the erosion of native genotypes or wild types endemic to this region. Newer areas will become suitable for cultivation.
Fennel and fenugreek	Plants under flowering stage at the time of frost suffer maximum yield loss.
Jasmine	Low temperatures may lead to shut down of flowering and reduction in bud size.
Carnation	Low Night temperature (<13°C) causes deterioration of flower quality due to calyx splitting.

References

Akhila, G. Nair and Annadurai, R. (2018). *International Journal of Pure and Applied Mathematics*, **119(14)**: 589–595.

Beniston, M. (2003). Climatic changes in mountain regions: a review of possible impacts. *Climatic change,* **59:** 5–31.

EIA. (2004). Greenhouse Gases' Effect on the Climate, Available at www.eia.doe.gov/oiaf/1605/ggccebro/chapter1.html (accessed November 2004).

FAO, (1995). FAOCLIM 1.2, worldwide agro climatic data. FAO Agrometeorology Working Paper Series No. 11. FAO, Rome. 1 CD-ROM and 66 p.

Grabherr G, Gottfried M.and Pauli, H. (1994). Climate effects on mountain plants. *Nature*, **369**: 448.

IPCC. (2007). Available at, www.ipcc.ch pdf assessment-report ar4 wg1 ar4-wg1-faqs.pdf.

Katz, R.W. and Brown, B.G. (1992). Extreme events in a changing climate: variability is more important than averages. *Climate Change,* **21**: 289-302.

La Sorte FA, Jetz, W. (2010). Projected range contractions of montane biodiversity under global warming. Proceedings of the Royal Society B: *Biological Sciences*, **277:** 3401–3410.

NASA. (2018). The relentless rise of carbon dioxide. https://climate.nasa.Gov/climate_resources.

Petit, J. R., Jouzel,J., Raynaud,D., Barkov,N. I., Barnola,J.-M. *et al.* (1999). Climate and atmospheric history of the past 420,000 years from the Vostok ice core, Antarctica, *Nature,* **399 (3):** 429–436. Available at www.nature.com

Petit-Maire, N. (1992). Lire l'avenir dans les archives géologiques. *La recherché*, **23 (243):** 566–569.

Pielke, R. A. Jr., and Landsea, C. W. (1998) Normalized hurricane damages in the United States: 1925–95. *Weather Forecasting*, **13:** 621–631.

Runeckles, V. C. and Krupa, S. V. (1994). The impact of UV B radiation and ozone on terrestrial vegetation. *Envir. Pollut.* **83(1-2):** 191–213.

Seitz, F. (1994). Global Warming and Ozone Hole Controversies, A Challenge to Scientific Judgement. G.C. Marshall Institute, Washington DC. 32 p.

Swaminathan, M.S. & Kesavan, P.C. (2012). Agricultural Research in an era of climate change. Agricultural Research, **12(1);** 3–11.

Tans, P.P., Fung, I.Y. and Takahashi, T. (1990). Observational constraints on the global atmospheric CO_2 budget. *Science,* **247:** 1431–1438.

Tirado, R. and Cotter, J. (2010). Ecological Farming: Drought-resistant Agriculture. Greenpeace Research Laboratories, University of Exeter, United Kingdom.

UNDEERC. (2018). Global Carbon Cycle. https://www.undeerc.org/pcor/sequestration/cycle.aspx (accessed December 2018).

Venkataraman, S. (2006). How is climate change affecting crop pests and diseases? Down to earth. Available at: https://www.downtoearth.org.in/blog/climate-change/how-is-climate-change-affecting-crop-pest-and-diseases--54199. Retrived on 16.02.2019.

Wim G. S. and Rene G. (1996). The climate change - Agriculture conundrum. Ed. Fakhri A. Bazzaz, Wim G. Sombroek, Global Climate Change and Agricultural Production. Published by John Wiley & Sons Ltd. England. pp.1–10.

Wolfe, D.W. and Erickson, J.D. (1993). Carbon dioxide effects on plants: uncertainties and implications for modelling crop response to climate change. In: Agricultural Dimensions of Global Climate Change. H.M. Kaiser and T.E. Drennen (eds.). pp. 153–178.

11

Vetiver-A Blessing to Coastal Ecosystem for an Integral Prosperity and Ecological Stability

S. Babu, S. Ramesh Kumar and M. Prakash
Department of Agronomy, Faculty of Agriculture, Annamalai University
Department of Horticulture, Faculty of Agriculture, Annamalai University
Department of Genetics and Plant Breeding, Faculty of Agriculture, Annamalai University

Introduction

Vetiver, commonly known as Khus grass is a perennial grass of Indian origin. Vetiver roots contain fragrant essential oil, which is a perfume by itself. Aroma chemicals such as vetriverol, vetriverone and vetriveryl acetate are prepared from this volatile oil. In India, it is mainly used in perfumes, cosmetics, and aromatherapy, food and flavouring industries. Since the plant has extensive finely structured fibrous roots, it is useful in both soil and water conservation and the plant itself is drought tolerant. The world production of vetiver oil is around 300 tons per annum of which India contributes about 20–25 tons only. The world major producers are Haiti, India, Java and Reunion. In India, it is cultivated in the states of Rajasthan, Uttar Pradesh, Karnataka, Tamil Nadu, Kerala, Andhra Pradesh and Telangana, with an annual production of about 20 tons of oil. The present consumption of vetiver oil in India is about 100 tons and 80% of the domestic consumption is met by export only. As the internal demand for vetiver oil is very high, concerns are risingover the improved production and quality of raw materials used. Vetiver is a miracle crop and a big boon for the farmers to increase incomes. This would increase farmer's income by three to four times. This crop has demand not only in India but also across the world. India is gaining big ground in cultivation of vetiver.

Basic information obout vetiver

Scientific name: Vetiveria zizanoides (Linn) Family: Poaceae

Local name: Usirah, Usira, Vira (Sanskrit), Khas, Khus(Hindi); Valo (Gujarati); Khas-khas (Bengali); Ramacham(Malayalam); Illamichamber (Tamil); Vattiveru (Telugu);Panni (Punjabi); Vala (Marathi); Khas (Urdu).

Plant Part Employed in Aromatic oil Extraction: The commercial, essential oil of vetiver is obtained by distillation of the roots. Vetiver is tolerant of extremes of temperature, soil moisture, and soil acidity and alkalinity (pH from 3.3 to 10.5) (Dalton, Smith, and Truong 1996).

Vetiver hedges, when fully established, provided adequate protection from floodwater over the 90-meter spacing on 0.2 to 0.35 percent land slope (Dalton, Smith, and Truong 1996). The vetiver hedges successfully reduced flood velocity and limited soil movement, resulting in little erosion in fallow strips and the complete protection of a young sorghum crop from flood damage (Dalton, Smith, and Truong 1996). Effectiveness of vetiver grass hedgerows in reducing flood damage to cultivated lands has already been reported (Dalton 1997). Vetiver system has been successfully applied as (1) backslopes, (2) sideslopes, (3) stream banks along the highway, (4) ditch linings, (5) shoulder slopes, (6) gabion walls, and (7) drainage structures in the highway situations (Sanguankaeo *et al.*, 2010).

Vetiver is highly tolerant to heavy metals. Vetiver root could accumulate higher heavy metal concentra-tions than shoot (Roongtanakiat *et al.*, 2007). However, an extremely high concentration of heavy metals could inhibit the growth of vetiver as shown by Roongtanakiat (2009). Vetiver grass (*Vetiveria zizanioides*) has been identified as the potential plant to be used in the phytoremediation treatment (Sharifah Nur Munirah Syed Hasan *et al.*, 2016). Vetiver is a major industrial crop and is grown for its essential oil that is used extensively world over in flavor and fragrance industries. Vetiver is also used in manufacture of handy-crafts, thatching houses, and organic compost production. Vetiver has been traditionally incorporated in the cropping systems of the region. (Prakasa Rao *et al.*, 2015).

Characteristics of the Plant

Vetiver (*Vetiveria zizanoides* (Linn) Nash.) or Khus of family Poaceae, is a perennial grass which can grow up to 1 to 2 meters hight and form wide clumps. The stems are erect and stiff and the leaves are 120–150 cm long, 0.8 cm wide and rather rigid. The panicles are 15–30 cm long, brownish-purple coloured and have 2.5–5.0 cm long branches. The spikelets are in pairs, and there are three stamens. The root system of vetiver is finely structured and very strong. It has no stolons or rhizomes. Unlike most grasses, which form horizontally spreading mat-like root systems, vetiver's fibrous roots grow downward, 2–4 m in depth, and are strongly scented. Vetiver is mainly cultivated for the fragrant essential oil distilled from its roots. The main chemical components of the oilare benzoic acid, vetiverol, furfurol, vetivone and vetivene. Due to its excellent fixative properties, it is used widely in perfumes. Dry roots are also used for making mats, fans, screens, pillows, baskets, incense sticks and sachet bags.

Major areas of cultivation

Vetiver is indigenous to India, Pakistan, Bangladesh, Sri Lanka and Malaysia. The main producers are Tropical Asia, Africa, Australia, Haiti, Indonesia, Guatemala, India, China and Brazil. It is also cultivated in Indonesia, Malaysia, Philippines, Japan, Angola, Belgian Congo, Dominican Republic, Argentina, British Guiana, Jamaica, Mauritius and Honduras. Worldwide production is estimated to about 250 tons per annum. In India, it is seen growing wild throughout Punjab, Uttar Pradesh and Assam and coastal tracks of Tamil Nadu. It is systematically cultivated as a crop in the states of Rajasthan, Uttar Pradesh, Kerala, Karnataka, Madhya Pradesh, Andhra Pradesh and Telangana. Annually 20–25 tons of oil is produced in India. Uttar Pradesh produces the highest quantity of oil, mainly through wild sources. Vetiver oil produced in North India is of premium quality and fetches a very high price in international market.

Characteristics of strain (s) for cultivation

In India, two types of vetiver namely 'South Indian' and 'North Indian' are generally under cultivation. North Indian types yield superior quality oil but its rooting finds to be shallow, especially in damp ground. South Indian types are the cultivated types with a thicker stem, less branching roots and wider leaves. It is non-seeding type, high yielding both in terms of root bio mass and oil. It is reproduced by vegetative propagation and it is the type suitable for erosion control. Among South Indian types, Pusa Hybrid-7, Hybrid-8, CIMAPKS-2, Sugandha, KH-8, KH-40 and ODV-3 are the varieties available for commercial cultivation. Cultivars Dharini, Gulabi and Kesari released by CIMAP, Lucknow were developed by repeated selection of germplasm collections from different parts of India.

Package of practices

Soil

Vetiver can be grown on almost every kind of soil. However, light soils, should be avoided as the roots grown in this soil produce very low percentage of oil. Well drained sandy loam and red lateritic soils rich in organic matter are considered to be ideal for cultivation. It can also be cultivated in clay loam soil but it is better to avoid clayey soil. It can be grown in wide pH range even in saline and alkaline soils with a pH of 8.5 to 10. A flat site is acceptable, but watering must be monitored to avoid water logging, that will stunt the growth of young plantlets. Mature vetiver, however, thrives under waterlogged conditions. It can also absorb dissolved heavy metals from polluted water and can tolerate arsenic, cadmium, chromium, nickel, lead, mercury, selenium and zinc.

Climate

Vetiver is tolerant to a wide range of temperature ranging from -15 °C to +55 °C, depending on growing region. The optimal soil temperature for root growth is 25 °C. Root dormancy occurs when temperature goes below 5 °C. Under frosty conditions, shoots become dormant and purple, or even die, but the underground growing points survive and can regrow quickly if the conditions improve. Shading affects vetiver's growth, but partial shading is acceptable. It is tolerant to drought, flood, and submergence and grows luxuriantly in places having moderately humid climate with annual rainfall of 1000 to 2000 mm. It can also be grown as an irrigated crop in other suitable places with scanty rainfall.

Propagation

Vetiver can be propagated either by seeds or slips, but slips are commonly used. The cultivated accessions which are propagated through vegetative means show limited variation, whereas, seed propagation is used for breeding new varieties. In North Indian types, profuse seeding and natural regeneration occurs from self-sown seeds. Seed yield varies between 400–650 kg/ha. Freshly collected spikelet show dormancy and require an after-ripening period of about 3 months. Removal of caryopsis from enclosed husk facilitates germination. Dormancy can also be broken by treating the seeds with gibberlic acid or potassium nitrate. In South Indian types, most of the spikelets are not subjected to fertilization and seeds which sometimes produced are very thin and are having a short dormancy period. In these non-seeding types, slips are separated from clumps of previous crops with rhizome portion intact having 15–20 cm of aerial portion is used for propagation. Slips thus obtained should be kept moist and stored in shade. Dry leaves should be removed from slips before transplanting to avoid carry over of pests and diseases.

Planting time

The most suitable time for planting vetiver is June – August with the onset of monsoon. In South Indian conditions, where diurnal variation in temperature is not significant and monsoon sets in early, the optimum planting time is February-April.

Land preparation

Land is ploughed to a depth of 20–25 cm by 2–3 deep ploughings and removed the perennial weeds. Recommended dose of farm yard manure or compost and fertilizers are applied and mixed well with the soil. In sloppy areas, pits are taken across the contour.

Planting

The mother clumps can be divided into small pieces to give many numbers of slips. Slips are separated from the clump with the rhizome portion intact having 15–20 cm of the shoot portion. While planting slips fibrous roots and leaves should be trimmed off. Ensure planting of slips at the correct time. Slips from healthy and disease free clumps are planted during June-July with the onset of monsoon vertically about 10 cm deep at a spacing of 60 × 30 cm/60 × 45 cm/60 × 60 cm based on soil fertility status, climate and variety and also based on the irrigation facility. Plant population varies from 27,800 to 1,10,000 plants/ha. If irrigation facilities are available, it is better to plant during March-April, and frequent irrigation will be required. Late planting resulted in the production of coarse roots which yield inferior quality oil.

Manuring

Normally, fertilizer application for vetiver is not practiced in fertile soils. But, on poor soils, 10 tons of FYM along with 25–50 kg/ha each of N, P_2O_5 can be applied. Care should be taken to apply N in 2–3 split doses. N: P_2O_5:K_2O dose of 60:22.5:22.5 is recommended in Kerala. Application of 60 kg P_2O_5/ha is suggested for vetiver cultivation in Central Uttar Pradesh.

Intercropping

During the initial crop growth (70–90 days), crops like cowpea, black gram, green gram, cluster bean, pigeon pea, senna and sacred basil can be grown.

Irrigation

In the absence of rainfall, soil moisture status should be maintained by irrigation from planting to establishment. In the areas where rainfall is good, well distributed over the year and humidity is high, supplementary irrigation is not necessary. However, in dry areas about 8–10 irrigations will be required to get the optimum yield. Apply mulch to conserve soil moisture. Irrigation should be discontinued 7–10 days before harvesting.

Intercultural operation

In case of newly established crop, 2–3 weeding and earthing-up at an interval of one month are needed during initial period of plant growth. Once the crop is established, weeds are kept under check because of vetiver's thick and dense shoot cover. Aerial portion is trimmed at 20–30 cm above ground level thrice during the entire cropping period of two years. First trimming should be done at 4–5 months after planting, second during second year just before flowering and third in second year winter season, about one month before digging of roots.

Plant protection

Insect pests

Vetiver is a hardy crop and infestation by pests is not a serious concern. However, in dry areas termites are seen damaging the crop. Grubs of beetle *Phyllophaga serrata* have also been reportedly infesting vetiver roots. These can be controlled by broadcasting neem cake @ 5 t/ha before final ploughing. Stem borer, *Chilo* sp. and scale insects are also a threat in some places to the commercially grown vetiver. Remove the leaves and plants severely infested by scales and spraying with neem oil 5% also reduces scale infestation. Nematode infestation of roots by is also reported. To prevent nematode infestation caused by *Heterodera zeae*, use nematode free healthy mother stock. High organic matter content of the soil, hot water treatment and application of neem cake @ 5 t/ha are also found effective in controlling nematode.

Diseases

During rainy season, the plant is infested by *Fusarium* sp. Leaf blight caused by *Curvularia trifolii* is another important disease during rainy season. The infested leaves bear tan to dark spots which turn black with age. The roots of affected plants become yellow and gradually dry out. These pathogens can be controlled by 2–3 spraying or drenching of copper oxy chloride 0.3%.

Harvesting

The time of harvesting of vetiver roots is very important as the yield of roots and oil percentage vary with changes in environmental conditions. Roots are harvested after 15–24 months of planting, but to obtain good quality oil it should be harvested at 18 months. Though,early harvesting gives higher essential oil yield, oil will be of lows specific gravity which also lack valuable high boiling constituents. If roots stay in ground for over two years, oil quality improves butyield diminishes considerably. Crop is generally harvested during

December - February by digging out the clumps along with its roots manually. A tractor drawn mould board plough can also be used for digging out roots up to 35 cm depth. Mechanical harvesting gives15% higher roots recovery over manual harvesting.

Processing

The harvested roots are separated from the aerial parts, washed thoroughly, chopped to shorter lengths of 5–10 cm to facilitate easy drying and then dried under shade for 1–2 days before distillation,which improves the olfactory quality of the essential oil, while prolonged sun drying reduces the oil yield. While drying, roots should be laid out in thin layers and this will prevent the chances of fungal growth that results in decomposition of root. Do not dry the roots on the ground in direct sunlight without close supervision as direct sunlight involves a high risk of degradation of its active principles. After drying, the oil is extracted from the roots through hydro or steam distillation. In North Indian varieties, distillation process is completed in 12–14 hours, while South Indian varieties require along duration of 72–96 hours, as it has low volatile oil and high boiling point. Two distinct fractions, one lighter than water and another heavier than water are obtained from vetiver. Heavier the oil better is the quality. After distillation is completed these fractions should be collected separately and later mixed together. The oil is then decanted and filtered. The distilled oil is treated with an hydrous sodium sulphate or common salt at the rate of 20 g/litre to remove the moisture. Oil obtained from stored roots is more viscous and possess a slightly better aroma than that obtained from freshly harvested roots. Fresh roots require less time for distillation and gives maximum oil yield.

The vetiver oil is amber brown and rather thick. Its odour is described as deep, sweet, woody, smoky, earthy, amber and balsam. Ageing of the essential oil for a period of six months improves the odour of the oil substantially wherein, the `harsh' `green' and `earthy odour' characters of the freshly distilled oil gets converted in to a fuller, heavier and sweeter odour. The oil should be stored in sealed amber coloured glass bottles or containers made of stainless steel, galvanised tanks, aluminium containers and stored in a cool and dry place.

Expected yield

The essential oil yield of vetiver roots varies considerably and it depends on a number of factors such as soil conditions, age of the roots, harvesting time, drying and distillation methods followed etc. On an average, the root yield may range from 3–4 tonnes per hectare from a two year old plantation. In sandy and sandy loam soils,root yield is as high as 2–2.5 tonnes/ha whereas; in salt affected areas only 1–1.5 tonnes of roots can be harvested per hectare. The average oil recovery from north Indian variety is between 0.15–0.2%, whereas, it is 1% from South Indian variety. Oil recovery from fresh roots is 0.3–0.8% and from dried roots it is 0.5–3.0%depending upon the duration of distillation. On an average, the oil recovery is around 1% on dry weight basis and 10–30 kg oil is obtained per hectare per crop.

Quality evaluation of essential oil

The vetiver oil should have the following specifications

Parameters	South Indian	North Indian
Colour	Brown	Reddish green
Specific gravity at 30 °C	0.990–1.015	1.512–1.523
Optical rotation	+10° to +25°	–50° to –132°
Refractive index at 30 °C	1.516–1.530	1.512–1.523
Saponification value	25–50	25–80
Saponification value after acetylated oil	125–155	145–200

Gas Liquid Chromatography composition of vetiver oil

Vetiver oil or khus oil is a complex oil containing over 100 identified components, mainly sesquiterpenes. The characteristic constituents were veteverol (45–80%), ß-vetispirene (1.6-4.5%), khusimol (3.4–13.7%), vetiselinenol (1.3–7.8%) and vetivone (2.5–6.3%). Besides these components, trace amounts of benzoic acid, vetivene, furfurol, khusemene, khusimone, ß-humulene, valencene, ß-vetivone, selinene etc. are also present in the oil.

Characteristic features of different cultivated varieties/strains

Variety	Characteristics
KS-1	Essential oil yield 17.8 kg/ha
Sugandha	Yields 21.2 q fresh roots/ha, 1.4% oil content and 19.7 kgha oil yield/ha.
Hybrid-8	Root yield 12–15 q/ha, 1% essential oil content and possess 70–85% vetiverol.
Keshari	Saffron flavoured, oil yield 30 kg/ha.
Gulabi	Has a rosy odour, tolerant to sodic soil, yields 2.8 t dry roots and 25–30 kg/ha of essential oil, can be cultivated in marginal soils and waste lands.
Dharini	Possesses longer, thicker and dense roots, tolerant to sodic soil, is a good soil binder and useful for soil and water conservation, oil yield 39 kg/ha.

Cultivation calendar

Major activity	Month	Activity details
Land preparation	May–June	2–3 deep ploughings & removal of perennial weeds.
Manure and fertilizer application	May–June	Application of basal dose of recommended dose of FYM/compost and fertilizers.

Major activity	Month	Activity details
Plantation	June–July	Slips from healthy, disease free clumps with rhizome portion intact having 15–20 cm of aerial portion are planted at a spacing of 60 × 30 cm/60 × 45 cm/60 × 60cm.
Irrigation	June–July	Irrigation should be given immediately after transplanting and up to establishment. Later on 8–10 irrigations are required throughout the cropping period.
Fertilizer application	July–August	Application of first top dressing of nitrogen 25 kg/ha at one month after planting
Inter cultural operations	July–August	2–3 weeding and earthing up at an interval of one month during initial period of establishment.
Inter cultural operations	October–November	Trimming of aerial portion at an height of 20–30 cm above ground 4–5 months after planting.
Inter cultural operations	March–April	Trimming of aerial portion at an height of 20–30 cm above ground just before flowering.
Fertilizer application	March–April	Application of second topdressing of nitrogen 25 kg/ha.
Intercultural operations	October–November	Trimming of aerial portion at an height of 20–30 cm above ground one month before digging.
Harvesting	December–February	Digging the clumps along with Its roots at eighteen months after planting either by manual or mechanical means.

Uses of Vetiver

Vetiver grass is grown for many purposes. The plant helps to stabilize soil and protects it against erosion, but it can also protect fields against pests and weeds. Vetiver has favorable qualities for animal feed. From the roots, oil is extracted and used for cosmetics, aromatherapy, herbal skincare and ayurvedic soap. Due to its fibrous properties, the plant can also be used for handicrafts, ropes and more. Since it has extensive fibrous roots, it is useful in both soil and water conservation. It helps in maintaining soil moisture, absorbs toxic substances in chemical fertilizers and pesticides and improves physical characteristics of soil. The plant is one of the best soil binders and is used in tropics to check soil erosion by planting along the contour. It is also widely grown as protective partitions in terraced fields and as a border for roads and gardens.

Skin care

Vetiver has been used to produce perfumes, creams and soaps. It is used for its antiseptic properties to treat acne and sores.

Soil and water conservation

Erosion control

Several aspects of vetiver make it an excellent erosion control plant in warmer climates. Vetiver's roots grow almost exclusively downward, 2 m (7 ft) to 4 m (13 ft), which is deeper than some tree roots. This makes vetiver an excellent stabilizing hedge for stream banks, terraces and rice paddies, and protects soil from sheet erosion. The roots bind to the soil, therefore it cannot dislodge. Vetiver has been used to stabilize railway cuttings/embankments in geologically challenging situations in an attempt to prevent mudslides and rock falls, such as the Konkan railway in western India. The plant also penetrates and loosens compacted soils.

Runoff mitigation and water conservation

The close-growing culms help to block surface water runoff. It slows the water flow and increases the amount absorbed by the soil (infiltration). It can withstand water velocity up to 5 m per second (16 ft/s). Vetiver mulch increases water infiltration and reduces evaporation, thus protects soil moisture under hot and dry conditions. The mulch also protects against splash erosion. In west African regions, such as Mali and Senegal, vetiver roots were traditionally used to reduce bacteria proliferation in water jugs and jars. In Indonesia, the roots of vetiver are widely used in the production of fragrant mats. In the Philippines and India, the roots are woven to make fragrant-smelling fans called "sandal root fans".

Crop protection and pest repellent

Vetiver can be used for crop protection. It attracts the stem borer (*Chilo partellus*), which lay their eggs preferentially on vetiver. Due to the hairy architecture of vetiver, the larvae cannot move on the leaves, fall to the ground and die. Vetiver's essential oil has anti-fungal properties against *Rhizoctonia solani*. As mulch, vetiver is used for weed control in coffee, cocoa and tea plantations. It builds a barrier in the form of a thick mat. When the mulch breaks down, soil organic matter is built up and additional crop nutrients become available.

Vetiver as a termite repellent

Vetiver extracts can repel termites. However, vetiver grass alone, unlike its extracts, cannot be used to repel termites. Unless the roots are damaged, the anti-termite chemicals, such as nootkatone, are not released.

Animal feed

The leaves of vetiver are a useful byproduct to feed cattle, goats, sheep and horses. The nutritional content depends on season, growth stage and soil fertility. Under moist climates, nutritional values and yields are best if vetiver is cut every 1–3 months.

Food and flavorings

Vetiver (Khus) is also used as a flavoring agent, usually as khus syrup. Khus syrup is made by adding khus essence to sugar, water and citric acid syrup. Khus essence is a dark green thick syrup made from the roots. It has a woody taste and a scent characteristic of khus. The syrup is used to flavor milkshakes and yogurt drinks like lassi, but can be used in ice creams, mixed beverages such as Shirley Temples and as a dessert topping. Khus syrup does not need to be refrigerated, although khus flavored products may need to be.

Perfumery and aromatherapy

Vetiver is mainly cultivated for the fragrant essential oil distilled from its roots. In perfumery, the older French spelling, vetiver, is often used. Worldwide production is estimated at about 250 tons per annum. Due to its excellent fixative properties, vetiver is used widely in perfumes. It is contained in 90% of all western perfumes. Vetiver is a more common ingredient in fragrances for men; some notable examples include Dior's Eau Sauvage, Guerlain Vétiver, M. Vétiver by Une Nuit à Bali,Zizan by Ormonde Jayne and Vétiver by L'Occitane en Provence

Essential oil

The oil is amber brown and viscous. Its odor is described as deep, sweet, woody, smoky, earthy, amber and balsam. Like patchouli and sandalwood essential oils, vetiver's odor develops and improves with aging. The oil's characteristics can vary significantly depending on where the grass is grown and the climate and soil conditions. The oil distilled in Haiti and Réunion has a more floral quality and is considered of higher quality than the smokier oil from Java. In north India, oil is distilled from wild-growing vetiver. This oil is known as khus or khas, and in India is considered superior to the oil obtained from the cultivated variety. It is rarely found outside of India, as most of it is consumed within the country.

Medicine

Vetiver has been used in traditional medicine in South Asia (India, Pakistan, Sri Lanka), Southeast Asia (Malaysia, Indonesia, Thailand), and West Africa. Old Tamil literature mentions the use of vetiver for medical purposes.

In-house

In the Indian Subcontinent, khus (vetiver roots) is often used to replace the straw or wood shaving pads in evaporative coolers. When cool water runs for months over wood shavings in evaporative cooler padding, they tend to accumulate algae, bacteria and other microorganisms. This causes the cooler to emit a fishy or seaweed smell. Vetiver root padding counteracts this odor. A cheaper alternative is to add vetiver cooler perfume or even pure khus attar to the tank. Another advantage is that vetiver padding does not catch fire as easily as dry wood shavings. Mats made by weaving vetiver roots and binding them with ropes or cords are used in India to cool rooms in a house during summer. The mats are typically hung in a doorway and kept moist by spraying with water periodically; they cool the passing air, as well as emitting a fresh aroma. In the hot summer months in India, sometimes a muslin sachet of vetiver roots is tossed into the earthen pot that keeps a household's drinking water cool. Like a bouquet garni, the bundle lends distinctive flavor and aroma to the water. Khus-scented syrups are also sold.

Soil remediation

A recent study found the plant is capable of growing in fuel-contaminated soil. In addition, the study discovered the plant is also able to clean the soil, so in the end, it is almost fuel-free.

Other uses

Vetiver grass is used as roof thatch (it lasts longer than other materials) and in mud brick-making for housing construction (such bricks have lower thermal conductivity). It is also made into strings and ropes, and grown as an ornamental plant (for the light purple flowers). Garlands made of vetiver grass are used to adorn the murti of Lord Nataraja (Shiva) in Hindu temples. It is a favourite offering to Ganesha. Vetiver oil has been used in an effort to track where mosquitoes live during dry seasons in Sub-Saharan Africa. Mosquitoes were tagged with strings soaked in vetiver oil then released. Dogs trained to track the scent, not native to Africa, found the marked mosquitoes in such places as holes in trees and in old termite mounds.

References

Dalton, P. A. (1997). Application of the vetiver grass hedges to erosion control on the croppedf lood plain of the Darling Downs. Master of Engineering thesis, University of Southern Queensland, Toowoomba, Queensland, Australia.

Dalton, P. A., Smith,R. J. and Truong, P. N. V.(1996). Vetiver grass hedges for erosion control on a cropped flood plain: Hedge hydraulics. *Agric. Water Management.* **31**; 91–104.

Sanguankaeo, S., Sawasdimongkol, L., and Jirawanwasana, P. (2010). Sustainable vetiver system in erosion control and stabilization for highways slopes in Thailand. Paper presented at the First Latin America Vetiver Conference, Santiago, Chile, 14-16 October (2010).

Roongtanakiat, N. (2009). Vetiver phytoremediation for heavy metal decontamination. TB 2009/1, Pacific Rim Vetiver Network, Office of the Royal Projects Development Board, Bangkok, Thailand.

Roongtanakiat, N., Tangruangkiat, A., and Meesat, R. (2007). Utilization of vetiver grass (*Vetiveria zizanioides*) removal of heavy metals from industrial waste waters. *Science Asia,* **33:** 397-403.

Sharifah Nur Munirah Syed Hasan, Faradiella Mohd Kusin, Ashton Lim Sue Lee, Tony Austin Ukang, Ferdaus Mohamat Yusuff, and Zelina Zaiton Ibrahim. Performance of Vetiver Grass (*Vetiveria zizanioides*) for Phytoremediation of Contaminated Water. MATEC Web of Conferences 103, 06003 (2017) DOI: 10.1051/matecconf/20171030 ISCEE 2016 6003.

Prakasa Rao, E.V.S. , Srinivas, Akshata & Gopinath, C.T. & Ravindra, N.s & Hebbar, Aparna & Prasad, H.N.. (2015). Vetiver Production for Small Farmers in India. 10.1007/978-3-319-16742-8_10.

12

Climate Change and Insect Pest Management in Coastal Agriculture Facts and Problems

S. Arivudainambi and V. Suhasini
Department of Entomology, Faculty of Agriculture, Annamalai University

Introduction

Global warming is a phenomenon of climate change characterized by a general increase in average temperatures of the Earth, which modifies the weather balances and ecosystems for a long time. In fact, the averages temperature of the planet has increased by 0.8° Celsius (33.4°F) compared to the end of the 19th century. Each of the last three decades has been warmer than all previous decades since the beginning of the statistical surveys in 1850. At the pace of current CO_2 emissions, scientists expect an increase of between 1.5° and 5.3°C (34.7° to 41.5°F) in average temperature by 2100 (Solar impulse foundation, 2018). Most of the several dozens of predictive models indicate that average temperature can increase by 1.7-5.3°C, as a result of doubling CO_2 concentration within next 60–100 years. 2.3°C is a value most often mentioned what means an increase by 0.3°C a decade.

Climate change has emerged as the most pressing global challenge of the 21st century and which is one of the largest and most complex problems, the developed community has ever faced. The impacts of higher temperatures, variable precipitation, and extreme weather events have already begun to impact the economic performance of countries and the lives and livelihoods of millions of poor people. India is among the countries most vulnerable to climate change. It has one of the highest densities of economic activity in the world, and very large numbers of poor people who rely on the natural resource base for their livelihoods, with a high dependence on rainfall. *By 2020, pressure on India's water, air, soil, and forests is expected to become the highest in the world.*

During the last few decades, the global agricultural production has risen and technology enhancement is still contributing to yield growth. However, population growth, water crisis, deforestation and climate change threatens the global food security.

Globally, about 2 billion people suffer from micronutrient deficiencies linked to lack of vegetable and legume consumption, while worldwide per capita consumption of vegetables and fruits is between 20 and 50 percent below recommended levels. Climate change will increase the risk of simultaneous crop failures across the world's biggest corn-growing regions and lead to less of the nutritionally critical vegetables that health experts say people aren't getting enough of already, scientists warn. "Vegetables and legumes are the vital components of a healthy, balanced and sustainable diet and nutritional guidelines consistently advise people to incorporate more vegetables and legumes into their diet."

Rising temperature affects flowering and leads to pests and disease buildup. Flood and excess rain over a short duration of time cause extensive damage to crops. Extreme weather events have caught attention of agrarian experts and scientists alike and they are now focussing on natural farming to arrest the impacts of climate change. By 2030, rice and wheat are likely to see about 6-10 per cent decrease in yields.

Despite technological advances, such as improved varieties, genetically modified organisms, and irrigation systems, weather is still a key factor in agricultural productivity, as well as soil properties and natural communities. The effect of climate on agriculture is related to variabilities in local climates rather than in global climate patterns. The Earth's average surface temperature has increased by 1.5 °F (0.83 °C) since 1880. A 2008 study published in *Science* suggested that, due to climate change, "Southern Africa could lose more than 30% of its main crop, maize, by 2030. In South Asia, losses of many regional staples, such as rice, millet and maize could top 10%.

In the long run, the climatic change could affect agriculture in several ways:

- *Productivity*, in terms of quantity and quality of crops.
- *Agricultural practices*, through changes of water use (irrigation) and agricultural inputs such as herbicides, insecticides and fertilizers.
- *Environmental effects*, in particular in relation of frequency and intensity of soil drainage (leading to nitrogen leaching), soil erosion, reduction of crop diversity.
- *Rural space*, through the loss and gain of cultivated lands, land speculation, land renunciation, and hydraulic amenities.
- *Adaptation*, organisms may become more or less competitive, as well as humans may develop urgency to develop more competitive organisms, such as flood resistant or salt resistant varieties of rice.

This article, deal with a comprehensive analysis related to yield loss, crop diversification, pest shift and problems with management models.

Crop Losses by Insect Pests

Since the beginning of agriculture, about 10,000 years ago, farmers compete with harmful organisms – animal pests (insects, mites, nematodes, rodents, slugs and snails, birds), plant pathogens (viruses, bacteria, fungi) and weeds (i.e. competitive plants) - collectively called as crop pests. Since crop production technology and especially crop protection methods are changing, loss data for major food and cash crops have been updated for the period 2001–03 (CABI's Crop Protection Compendium, 2018). Crop losses may be quantitative and/ or qualitative. Quantitative losses result from reduced productivity, leading to a less yield per unit area. Qualitative losses from pests may result from the reduced content of valuable ingredients, reduced market quality, reduced storage characteristics or due to the contamination of the harvested product with pests, parts of pests or toxic products of the pests.

Insect pests have potential to reduce crop production substantially. About 30-35% of the annual crop yield in India gets wasted because of pests, nematodes. Without preventive protection with pesticides, natural enemies, host plant resistance and other non-chemical controls, 70% of crops could have been lost to pests (Oerke, 2006). Despite pesticide application 15 to 25% yield loss was reported in many crops. The loss would be very high if were crops grown without spray. For example fruits, vegetables, cotton and cereals will face 78, 54, 83 and 32% yield loss respectively. In the year 2016–17, cotton area in Punjab & Haryana reduced up to 27% (7.56 lakh hectares) because of the loss happened by insect pests in the previous years.

Rising global temperatures increase loss due to insect pests; because insect pests devour far more of the world's crops, according to the first global analysis of the subject, even if climate change is restricted to the international target of 2°C. Dramatic fluctuations in conventional rain patterns and monsoonal systems continue to inflict great hardship on farmers.

Table 1. Percent yield loss in crops (various years)

Crop	1965	1990	2000	2015
Cotton	18	30	11	39
Rice	10	18	25	27
Pulses	5	8	15	15
Sugarcane	10	16	20	20
Average	**7.2**	**23.3**	**36.2**	**38.4**

The world's top four grain producers will lose far more to pests with just 2C of global warning

% of yield lost to field pests - top producers in red.

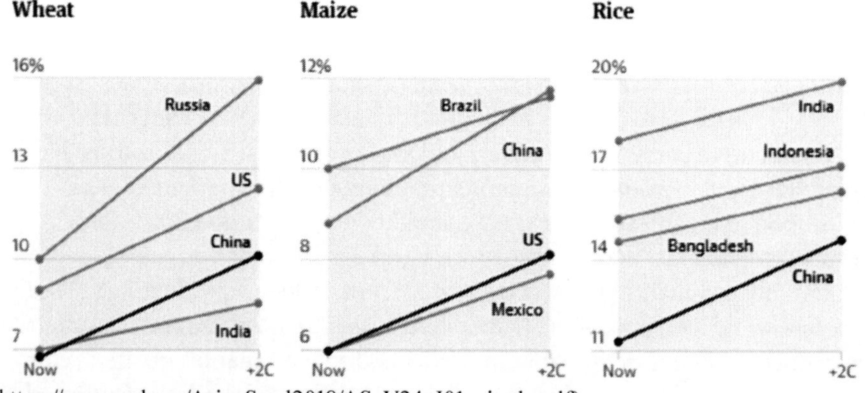

(https://apsaseed.org/AsianSeed2018/AS_V24_I01_single.pdf)

Coastal environment

Agricultural diversification is an important mechanism for economic growth. To meet the challenges of a globalizing market in agriculture as well as the growing and changing needs of the population, many countries have undertaken crop diversification to enhance productivity and cultivate high value crops with positive outcome. Indian states are gradually diversifying their crop sector in favour of high value commodities, especially fruits, vegetables and spices. There was a tendency towards cereal centric specialization. But, later when increased productivity of food grains, especially cereals, made it possible to allocate more area to other crops such as oilseeds with a severe supply shortage, the specialization tendency witnessed earlier has given room for overall crop diversification.

Coastal environment are highly fragile with less biodiversity and low fertility. Crop diversification is considered as a resilience mechanism followed by farmers in different regions. Socio-ecological systems of coastal areas are more vulnerable to the impact of climatic changes.

Palanisami *et al.* (2009) have examined the vulnerabilities of the coastal districts of Tamil Nadu to climatic change. They have concluded that Ramanathapuram and Nagapattinam districts are most vulnerable to climatic change. The crop diversification indices of the two districts for the year 2005–06 (Table 2) are respectively 0.403 and 0.344 which means that only about 40% of the agricultural area is occupied by diverse crops. This shows that there is an inverse relation between crop diversification and vulnerability to climatic change. Resilience in general refers to the level of resistance or recovery from shocks. In the case of coastal districts, the normal shock will be changes in rainfall resulting in floods or droughts, or other natural calamities. One of the major resistance identified was the cropping pattern changes in the coastal districts for the past 30 years. The area under food crops in coastal regions which was 88.8% during 1975–76 is reduced to 82.7% in 2005-06 and the percentage area under non food crops increased from 11.2% to 17.3%. This clearly confirms that farmers in the coastal region are resilient to climatic changes by changing cropping pattern. Hence it is confirmed that crop diversification is considered as one of the resilient mechanism particularly in the coastal region.

Table 2. Crop Diversification Indices (Modified Entrophy Index)

District/year	1980–81	1990–91	2000–01	2005–06	2010–11	2011–12
Kancheepuram	0.722	0.768	0.768	0.747	0.741	0.748
Cuddalore	0.634	0.717	0.623	0.622	0.421	0.411
Nagapattinam	0.414	0.469	0.397	0.344	0.321	0.320
Thanjavur	0.493	0.566	0.386	0.328	0.314	0.311
Pudukottai	0.449	0.574	0.425	0.453	0.455	0.452
Ramanathapuram	0.449	0.373	0.379	0.403	0.387	0.352
Tirunelveli	0.571	0.646	0.663	0.593	0.639	0.683
Thoothukudi	0.839	0.811	0.864	0.772	0.774	0.734
Kanniyakumari	0.319	0.408	0.509	0.516	0.579	0.611

(Kalai selvi *et al.*, 2012)

Wide spatio-temporal disparity in the diversification of crops is observed in the coastal districts of Tamil Nadu. The diversification in the coastal districts can be effectively utilized by strengthening of the linkage between agricultural and industrial sectors. In those districts where the deceleration of diversification had been exhibited, efforts must be taken for singling out the causative factors and adoption of appropriate measures for augmenting the diversification.

In this regard, the technology has a dominant role to play and as such adequate measures should be taken for propagating the innovative technologies in agriculture among the coastal farmers. Besides, diversification of enterprises should also be encouraged as a measure of minimizing the risk via resilience mechanism in those coastal districts where the index of diversification had showed plateau over years which will help to minimize the crop failure and income loss including employment to the rural people.

Yield loss

Yield losses in crops are certainly based on the interactions shown below.

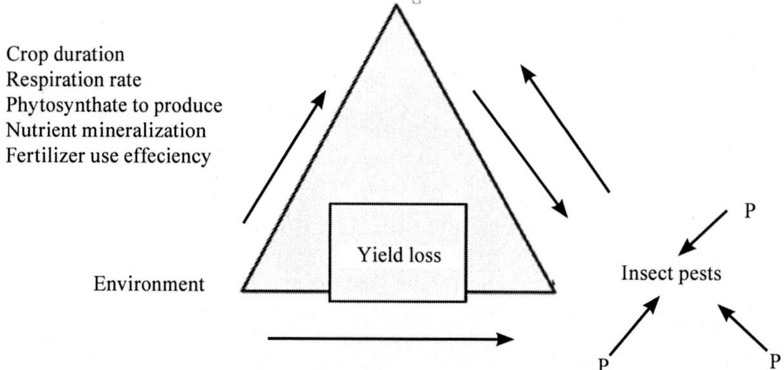

Direct effect of Environment on Pest Population

- Development & Development rates of Pests
- Dispersal of Pests
- Voltinism of Pests
- Density of Pests
- Composition of populations of Pests
- Pest outbreaks

Global warming could lead to an increase in pest insect populations, harming yields of staple crops like wheat, soybean, and corn. While warmer temperatures create longer growing seasons, and faster growth rates for plants, it also increases the metabolic rate and number of breeding cycles of insect populations. Insects that previously had only two breeding cycles per year could gain an additional cycle if warm growing seasons extend, causing a population boom. Temperate places and higher latitudes are more likely to experience a dramatic change in insect populations.

Temperature affects both the time of development as well as fecundity; consequently, the appearance and dynamics of insect populations in the field are dictated by ambient temperature. The effects of temperature on the melon thrips, *Thrips palmi* karny (Thysanoptera: Thripidae), preimaginal development, survival, fecundity, longevity of females and males and population growth were investigated indicating that the duration of egg, larval and pupal stages was significantly influenced by increased temperature. The egg to adult development period of *T. Palmi* declined from 35.7 to 9.6 days as the temperature increased from 16 to 31°C. The development threshold temperature estimated was 13.91, 11.82, 9.36 and 10.45°C for adult preoviposition period, total preoviposition period, female longevity, and male longevity, respectively. The thermal for competing the adult preovi position period, total preoviposition period, female longevity and male longevity were 29.3, 227.3, 454.6 and 344.8 degree days, respectively. Female longevity was found to be shortest at 31°C (18.7 days) and longest at 16°C(50.7 days). Fecundity was highest at 25°C (64.2 eggs/female) and lowest at 16°C (7.6). The optimal development temperature for *T. Palmi* in egg plant, *Solanum melongena* L. (Solanales: Solanaceae), ws determined to be 25°C (Ramchandra Yadav and Niann-Tai Chang, 2014).

The basic climate parameters, i.e. temperature and humidity, influence insects both directly and indirectly. The direct influence can be observed through limiting and stimulating the activity of larvae and adults, insects dispersal in the environment, phenology and growing length, as well as through the possibility of surviving in adverse weather conditions population genetics, etc. Indirect influence includes a climatic influence on environment where insects appear, such as influence on plant formations, plant phenology, food quality, predators, parasitoids and activity of entomopathogens.

Insects as poikilothermic animals change activity depending on the temperature of the surrounding environment (Bale *et al.*, 2002; Menéndez *et al.*, 2007). Increasing the temperature to the thermal optima level causes acceleration of the insect metabolism. Hence, it directly influences their increased activity. In the temperate climate zone conditions, the average temperature increase is followed by i.a. more intensive and longer total day and night's activity of *imago* of majority phytophagous species in forest environment, implied as feeding and mating, as well as time spent on finding proper place for laying eggs (Moore and Allard 2008; Netherer and Schopf 2010). It can also result in insects dispersion increase in the forest environment, as well as more frequent oviposition and possibility of colonising larger number of host plants (Parvatha Reddy, 2013).

In higher temperature conditions, the development of egg, larva and pupa shortens, which is the characteristic phenomenon for large group of forest species (Szujecki, 1998). Faster development of pre-imaginal stages implies shorter time of exposure to adverse environmental conditions such as low temperature, too high or insufficient humidity, attacks of predators and parasitoids, and entomopathogen's activity. It can result in reproductive success of many insect species. Temperature influence on a length of larval development has been observed under laboratorial conditions for two significant species of native foliophages: the nun moth, *Lymantria monacha* (L.) and the gypsy moth *Lymantria dispar* (L.) (Karolewski *et al.,* 2007). In both cases, temperature increase has had an influence on reducing growth period, from egg phase to pupa. Different results have been obtained regarding larva survivability of both species. When the average environment temperature has increased, higher mortality has been observed for caterpillars of *L. monacha*. Whereas the survivability of *L. dispar* larvae has increased. These differences probably result from two different thermal optima for both species reflected in varied environmental preferences. The results of the described experiment present variety of climatic parameters influence on the insect development, even when closely related species are compared.

Climate changes can cause faster evolutionary adaptation than usually. Menéndez *et al.* 2007), Moore and Allard (2008) and *Régnière* (*2009*) have presented a short review of researches on it regarding selected insect species. They have observed how European butterfly Brown Argus Aricia agestis (Den. & Schiff.) are adapted to new thermal conditions in short period of time by shifting diapause-inducing temperature threshold. Another example of such phenomenon is the chrysomelid beetle *Chrysomela aeneicollis* (Schaeff.) for which an increase in allele frequency is responsible for the synthesis of low-temperature proteins resistance has been observed. *Hill et al.* (*1999*) have presented results that show morphological changes in the population of the butterfly *Pararge aegeria* (L.). Individuals from the population that colonised new areas in Great Britain for 20 years before the studies have started had larger wing area surface as well as weight of the thorax in comparison with individuals from the settled populations. The increase in dispersal forms has also been observed on the British Isles for the two bush-cricket species *Conocephalus discolor* (F.) and *Metrioptera roeselii* (Hagenbach) that have extended their previous range (*Thomas et al., 2001*).

Temperature and humidity change can influence insects indirectly by changes in host plants metabolism and physiology (*Ayres, Lombardero 2000; Rouault et al.*, 2006; Moore and Allard, 2008; Netherer and Schopf, 2010). In general, it is indicated that long and intense droughts, have negative impact on plants' condition, thus increasing their susceptibility to phytophagous insects. Dying of oak stands, as a result of water shortage followed by folivore, cambiophage, and xylophage attacks, is a current example of such interaction observed in Europe (*Thomas, 2008*). Although a moderate temperature increase (as well as and CO_2concentration) can cause a decrease in food quality for some foliophages, as a result of nitrogen level decrease in foliage, as well as an increase in the synthesis of secondary metabolites, e.g. tannins (*Buse et al. 1998; Dury et al., 1998; Kuokkanen et al.*, 2001). It has an influence on deterioration of plant as food and may increase plant resistance. Huberthy and Denno (2004) and Rouault *et al.* (2006) have conducted a result meta-analysis regarding the influence of plant humidity shortage on development, survivability and fertility of phytophagous insects. Their studies have been inspired by observed discrepancy between outbreaks number in natural environment that often occur after droughts, and results that have showed negative influence of water shortage on phytophagous insects. Analysis results have indicated that reactions of phytophagous insects to water level decrease in plant tissues depend on their affinity to ecological guilds, so to the group of species sharing similar feeding habit. Positive influence of drought (especially long-lasting drought) has been observed with regard to insects developing in wood, whereas the decrease in water and turgor level in plant cells has had negative influence on species that suck out liquids from tissues (aphids) and on species that develop in galls. Analysis results of the influence on other phytophagous insects, external leaf feeders and leaf miners, have been ambiguous.

Additional life cycles/season

- More generations - crop damage/year
- Reproductive potential
- Faster resistance to insecticides
- Greater winter survival
- 1°C increase in temperature reduces winter mortality by about 16.5%.

Rising temperatures extend the growing season, greater nutrient demands coincide with planting and fruiting of many crops.

- The distribution of species has recently shifted to higher elevations at a median rate of 11.0 meters per decade, and to higher latitudes at a median rate of 16.9 kilometers per decade.
- Migration up elevation gradient
- Development & oviposition
- Out breaks,
- Invasive sp.
- Bio-control efficiency of entomopathogens
- Reliability of ETL
- Predation & parasitism
- Predator's searching efficiency - At 35°C the attacking rate decreased drastically.
- Direct effect
- Indirect effect

Natural enemies are another element of the ecosystem by which climatic changes influence indirectly phytophagous insects. Enemies activity and effectiveness and the way of influencing phytophagous populations can be diverse. Furthermore, interrelation of both the elements (phytophages versus natural enemies) is complicated by indirect climate influence on host plants. For instance, lower food quality of plants in result of drought causes longer development of phytophages, what determines higher probability of the attacks of natural enemies, such as predators and parasitoids (Coviella, Trumble 1999; Rouault *et al.*, 2006). On the other hand, plant chemism change, caused by climate parameters influence, results in quality change (size and chemical contents) of phytophagous insects as hosts, for example, of parasitoid larvae. It influences elements as parasitoid effectiveness of the victim search, the egg number, the size and sex ratio (Coviella, Trumble 1999). Higher temperature as the stimulating factor can cause activity increase of natural enemies and their faster development (Netherer and Schopf 2010, Sharma, 2014). Temperature influence can also regard phytophagous insects itself and as such it can influence their susceptibility to enemy attacks. For instance, under higher temperature conditions, weaken reaction for alarm pheromones, produced in case of predator or parasitoid attack, has been observed for aphids (Awmack *et al.,* 1997). On the other hand, faster development, induced by higher temperature, especially in instars exposed to parasitoid attacks, can result in higher survivability of some phytophages (Petzoldt, Seaman 2006).

Range shift of phytophagous insects

The current distribution pattern in most insect species is effect of climate. The phenomenon can be observed particularly on range borders where temperature is a main limiting factor. For instance, –16°C is the critical value for North American species of bark beetle *Dendroctonus frontalis*Zimm., which is one of the most dangerous pests for coniferous trees in the region. Nearly absolute mortality of the population occurs below this value. Such temperature is observed on the northern range border of the aforementioned species (Ayres, Lombardero, 2000).

It implies that average temperature increase can enable more termophilous species to expand in the northern direction and on higher altitudes. Simultaneously, southern and lower range borders can be shifted (Parmesan 1996; Walther *et al.*, 2002; Parmesan, Yohe 2003; Menéndez 2007; Battisti 2008). With regard to many phytophagous insects, a range increase is probable also because species' ranges are smaller than areas where their host plants grow. Many examples of insects' range shift were observed in recent years. In the 1990s, few leaf mining moths of the family Gracillariidae have occurred in the Central Europe, including Poland (Šefrová 2003). The horse-chestnut leaf miner Cameraria ohridella Deschka & Dimić that attacks horse chestnuts *Aesculus hippocastanum* was the most spectacular example among them. Apart from this accidental introduction of the pest, shifting of the northern and eastern range borders, as a result of temperature increase, was the most probable cause of the species expansion to new areas. Range shift of forest folivores in Europe has been well researched for two species of geometrid moths, Winter moth, *O. brumata*, and Autumnal moth, *Epirrita autumnata* (Borkh.), in forest stands of northern Scandinavia (Jepsen *et al.,* 2008). Cyclic outbreaks have been observed for both species in the aforementioned area, sometimes leading to substantial loss of foliage. For the last 15–20 years, areas of both defoliators' mass outbreaks have been increased significantly. *Operophtera brumata*, a species less resistant to low temperature, has expanded to the north-east to areas where *E. autumnata* was the dominant species so far. The latter has increased range to areas located inland and characterised by cooler climate. The Pine processionary moth *Thaumetopoea pityocampa* Den. & Schiff are another well-documented examples of the species range shift with regard to the influence on forest management. The species is recognised to be one of the main foliophagous pests in the Mediterranean region. Temperature in winter, when caterpillars feed on needles of various pine species (rarely on other coniferous species), is the main factor that influences range limits of *T. pityocampa*. From the mid-1970s to 2004, the species enlarged its range in France to the north direction by almost 90 km. In the same period, its upper range border in Italian Alps moved up by over 200 m in some regions (Battisti 2008; Battisti *et al.*, 2005). Same observations have been made for changes in an upper range border of the species in Spanish Sierra Nevada (Hódar, Zamora 2004). Average temperature increase enabled expansions to areas that have not been colonised before. Higher survivability of caterpillars in winter, during feeding time, was observed (Battisti *et al.*, 2005, Buffo *et al.*, 2007), whereas warmer nights in summer (with temperature over 14°C) influenced distance and altitude increase of female expansion (Battisti *et al.*, 2006). It has been often pointed out at the necessity of constant monitoring of insects range shifts and of selecting either species or groups of insect species that would indicate changes in forest environment (for instance, Ayres, Lombardero 2000; Bale *et al.*, 2002; Logan *et al.* 2003; Menéndez *et al.*, 2007). Attempts at predicting species range shift have also been made (for instance, Williams, Liebhold 1995 a,b; Jönsson *et al.* 2007; Régnière 2009). Apart from these ecological requirements of indicator organisms, varied factors are included in the aforementioned research, such as typology (for instance, the type of habitats and plant formations) and climate parameters (average, minimum and maximum temperature/precipitation per month). The variability of the latter implies noticeable bias in any attempts to predict changes in insects range. Hence, such predictions can only be seen as possible scenarios. Results can be influenced by relatively small changes in parameters that with regard to climate unpredictability (even in few-year scale) can hinder from making any exact prognosis. Williams and Liebhold (1995a, b) have conducted prognostic research on insect range shift. They have used data on defoliated forest stands in the states of Oregon and Pennsylvania that were previously exposed to attacks of tortricid moth *Choristoneura occidentalis* Free and of L. dispar. Alternative scenarios have been discussed include: (a) average temperature increase by 2°C and unchanged precipitation level, (b) average temperature increase by 2°C and precipitation level decrease by 0.5 mm/day and (c) average increase in values of both parameters. Average temperature increase and unchanged precipitation level have been factors that caused *L. dispar* expansion increases, while predicted range of *Ch. occidentalis* has decreased. The increasing assumed temperature and decreasing precipitation level have caused range decrease of both defoliators, whereas increase of both parameters was positively correlated with the growth of the outbreaks areas. Similar research has also been conducted in Finland (*Vanhanen et al., 2007*). Probable range shifts of *L. monachal* and *L. dispar* have been predicted on the basis of selected average temperature change scenarios that are included in Intergovernmental Panel on Climate Change (IPCC) report of 2001. Each simulation (*i.e.* temperature increase by 1.4, 3.6 and 5.8°C) has initiated range shift for both species. Northern and southern range borders of both species have shifted by 500–700 and 100–900 km, respectively, in the North Pole direction.

Cyclic outbreaks in some of phytophagous insects are connected with decrease forest condition decrease, losses in forest production and with bearing high costs of controlling pest population. Therefore, it is essential to answer the question to what degree predicated climate change will interrelate with negative influence of insect species on forests. American researchers have conducted interesting studies in this respect (*Currano et al., 2008*). They have analysed plant fossils dated for 59–52 million years (at the turn of the Paleocene and Eocene). One of the highest temperature increases was noted on Earth at this time (by about 6°C) in result of larger CO_2 concentration in the air. They have observed that the average level of damages in leaf lamina made by folivores and the average temperature increase, as well as CO_2 concentration, were positively correlated. It was explained by the increase in CO_2 concentration in the air which interrelates with increase in carbon-to-nitrogen ratio in plant tissues resulting in decrease in the nutritive values of the foliage. Hence, losses from nutritional value of lower leaves need to be compensated by higher consumption. Comparing the aforementioned results with the currently observed climate changes, increase in defoliation levels should be expected, followed by increased damage of forest stands (Rouault *et al.*, 2006; Battisti 2008; Currano *et al.*, 2008; DeLucia *et al.*, 2008). Changing thermal conditions and humidity both can have positive and negative influence on insects. Battisti (2008) has given an example of two forest phytophages on which temperature increase had significantly different influence. Between 1985 and 1992, an unexpected mass outbreak of web-spinning sawfly *Cephalcia arvensis* Panzer, an oligophagous hymenopteran species associated with spruce, was observed in Southern Alps. The species usually do not have a tendency to mass outbreaks, which result from limited dispersal abilities and low female fertility, as well as from long (even up to few years) diapause followed by higher mortality in the population. According to authors, the outbreak of *C. arvensis* was caused by few-year-period of high average temperatures and drought in June and July, during larval feeding time. On the one hand, higher temperature and low humidity caused shorter development of larvae of *C. arvensis;* on the other hand, enabled faster pupation and avoiding the longer diapause. In effect, the species has produced generation once a year for few years, which caused sudden eruption of population density. The opposite phenomenon has also been observed in the region of Alps for another folivore species, Larch Tortrix Zeiraphera griseana (=diniana). The species develops on few coniferous species; however, its main host plant in the discussed area is larch. The species is of great significance because of 8- to 10-year cycles of outbreaks; the history of which was estimated on the basis of dendrochronological research dating back over a thousand years (*Esper et al.*, 2006). Caterpillars of *Z. griseana* hatch in spring and commence feeding on developing needles. Starting from 1989, a significant decrease in the number of larch tortrix has been observed for few seasons. In effect, the number of outbreaks has dropped as well. Meteorological analyses have showed that high temperature in winter and spring influenced higher mortality of eggs and disturbance of synchronisation between eggs 'hatch and needles' development. Climate change can also cause higher activity of pest species that was not significant before for the forest management in the area. The outbreak frequency of the European pine sawfly, *Neodiprion sertifer* (Geoff.), regarding temperature increase and selected environmental elements, has been analysed in Finland (*Virtanen et al.*, 1996). Research has showed that days with the temperature lower than –36°C in winter is the main factor limiting *N. sertifer* outbreaks in the Northern Finland because the high mortality of eggs is observed below this temperature. The same research has focused on scenario of the average winter temperature increase by 3.6°C to 2050. According to authors, the global warming can cause increase in the outbreaks frequency of the European pine sawfly in areas where the species does not occur or occurs sporadically. Similar simulations regarding population dynamics of European spruce bark beetle have been performed in Southern Sweden (*Jönsson et al.,* 2007). Currently, the species has only one generation during the year in the researched area. By predicted increase in yearly average temperature by 2–3°C, the second swarming of the beetles is possible and increase by 5–6°C can cause development of the another generation. However, the authors have noticed that favourable thermal conditions for *I. typographus* can occur every year, even if the predicted scenario will be realised. Adequately, early time of spring swarming and shorter period of development preimaginal stages influence the possible development of the second generation of the species.

Invasive species Insects belong to the group of animals where alien species are the most frequently represented for the area. Climate change can result in adaptation, population increase and expansion of alien species that are better adapted than native taxa (Capdevila-Argüelles, Zilletti 2008). Temperature increase can positively influence the population number of introduced species whose development and survivability were limited by low temperature. Apart from climate change, human activity is the significant factor in the process, such as intentional or accidental introducing of exotic plants and phytophagous species. In European forest ecosystem, two moths in the family Gracillariidae are claimed to be invasive: *Parectopa robiniella* Clem. and*Phyllonorycter robiniella* (Clem.) that arrived to Europe from North America (*Šefrová,* 2003). Caterpillars of both species develop in leaf mines on the leaves of the black locust *Robinia pseudoacacia* L., a tree species introduced in Europe at the beginning of the 17th century. Even though the host plant has been present for the several centuries in Europe, both insect species were recorded for the first time in the second half of the 20th century in the Southern regions of the continent. Since then, the expansion process is observed in the Northern direction, and the phenomenon is usually explained to be caused by global warming.

Host shifts of phytophagous insects, Insect range shifts, resulting from changing climate parameters influence, can also cause insects adaptation to new host plants. The situation occurs especially when closely related species of the host plant exists in the new range of the phytophagous insect. Widening of host plants spectrum or even a change in feeding preferences by using available niches can occur. *Thaumetopoea pityocampa*, observed in Serra Nevada Mountains, in southeast Spain, is an example of such a phenomenon.

The average temperature increase for the last several decades enabled species dispersion to the higher altitudes where it has not existed before. Range shift was accompanied by adaptation to the new host plant, i.e. a relict subspecies of Scots pine, *Pinus sylvestris* var. nevadensis (*Hódar, Zamora* 2004).

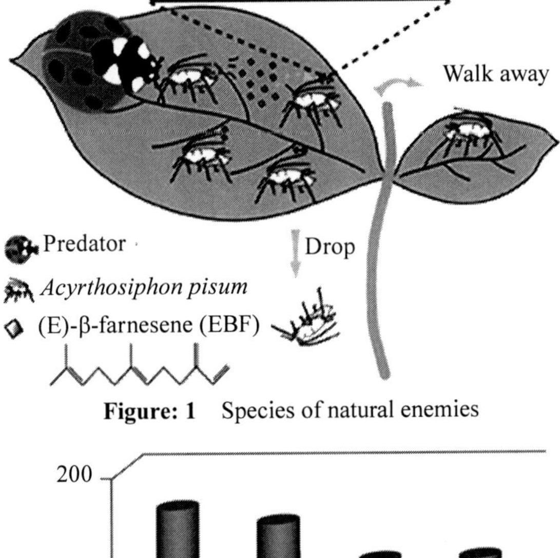

Figure: 1 Species of natural enemies

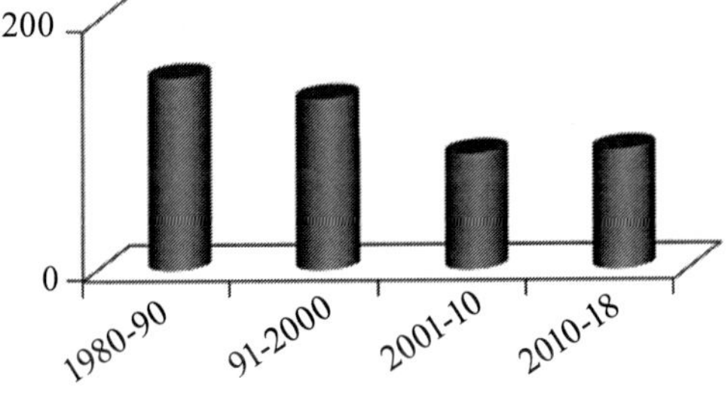

Figure: 2 Natural enemies population

Effects of elevated CO_2

The University of Illinois conducted studies to measure the effect of warmer temperatures on soybean plant growth and Japanese beetle populations. Warmer temperatures and elevated CO_2 levels were simulated for one field of soybeans, while the other was left as a control. These studies found that the soybeans with elevated CO_2 levels grew much faster and had higher yields, also attracted Japanese beetles at a significantly higher rate than the control field. The beetles in the field with increased CO_2 levels also laid more eggs on the soybean plants and had longer life spans, indicating the possibility of a rapidly expanding population. DeLucia projected that if the project were to continue, the field with elevated CO_2 levels would eventually show lower yields than that of the control field.

The increased CO_2 levels deactivated three genes within the soybean plant that normally create chemical defences against pest insects. One of these defences is a protein that blocks digestion of the soy leaves in insects. Since this gene was deactivated, the beetles were able to digest a much higher amount of plant matter than the beetles in the control field. This led to the observed longer life spans and higher egg-laying rates in the experimental field.

There are few proposed solutions to the issue of expanding pest populations. One proposed solution is to increase the number of pesticides used on future crops. This has the benefit of being relatively cost effective and simple, but may be ineffective. Many insects have been building up an immunity to these pesticides. Another proposed solution is to utilize biological control agents. This includes things like planting rows of native vegetation in between rows of crops. This solution is beneficial in its overall environmental impact. Not only are more native plants getting planted, but insects pest are no longer building up an immunity to pesticides. However, planting additional native plants requires more room, which destroys additional acres of public land. The cost is also much higher than simply using pesticides.

- Less nutritious to insect herbivores
- As a consequence, insects take more time converting the food they eat into biomass.
- Increased carbon to nitrogen ratios in plant tissue resulting from increased CO_2 levels may slow.
- In order to mitigate the effects of less nutritious food, insect herbivores often consume more.
- Soybeans grown in elevated CO_2 atmosphere had 59% more damage from insects.
- Decreased water content in leaf and
- Trichome density
- Increased feeding in mustard, cabbage and leaf mining.
- In India, the average CO_2 level was 399 ppm. However, at a coastal station in Goa, the level shot up to 408 ppm.

Elevated CO_2 on transgenic plants and insect pests			
Tobaco bud worm	*Spodoptra exigua*	Limited N produced less toxin	Coviella *et al.*, 2000
	Spodoptra exigua	N based toxin was affected	Coviella *et al.*, 2002
Boll worm	*H. armigera*	Decrease in Bt toxin level	Ge Feng *et al.*, 2005

Pest complex under elevated CO2		
Insects	**Elevated CO_2 (570 ppm)**	**References**
Japanese beetle Potato leaf hopprer Western corn root worm Mexican bean beetle	57 % more damage	Trumble and Butler, 2009
Thrips	90 % more feeding	Heagle, 2003
Cereal aphid	Higher population	Newman, 2004

Pest shift

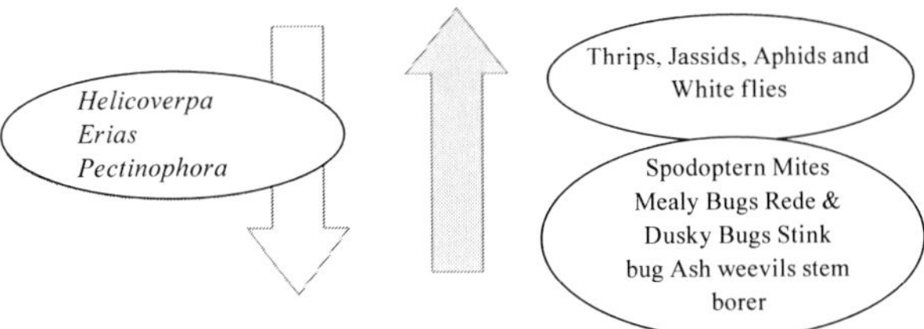

USA - 5 insecticide applications against lepidopteran insect pests Florida conditions require 15–32 applications .

Additionally, some classes of pesticides (pyrethroids and spinosad) have been shown to be less effective in controlling insects at higher temperatures

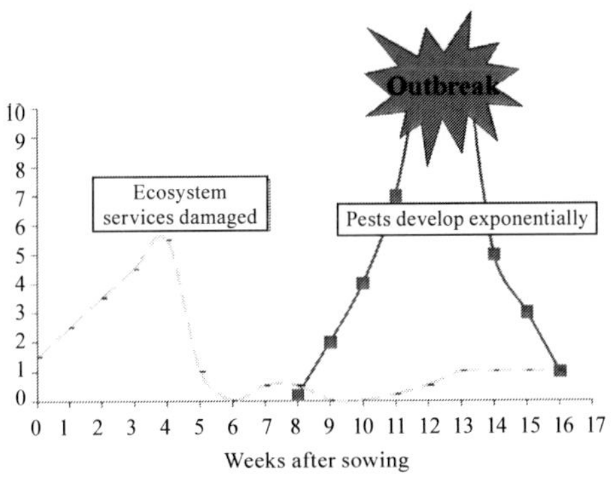

Period	Amount/ha – a.i.
2000-01	0.2
2002-03	0.2
2004-05	0.21
2006-07	0.21
2008-09	0.22
2010-11	0.25
2012-13	0.3
2014-15	0.4
2016-17	0.6

(*FAO*, 2017)

Challenges

- New pests
- Biotypes
- Density
- More damage
- Aphid, whitefly,
- Mealybug, scales,
- Mites, thrips

There are currently many evidences for climate change on Earth. The most popular hypothesis states that average temperature will increase in result of higher CO_2 concentration in the atmosphere, among others. Predicted changes will have a significant influence on forest ecosystems and on all elements of forest biocenosis. Climate changes are important for phytophagous insect species, and basic climate parameters, temperature and humidity influence it both directly and indirectly.

1. Global warming is conducive to polyphagous and eurytopic species. It results from higher ecological plasticity and adapting abilities of the organisms.
2. The general observations of the climate change influence on phytophagous insects suggest that the role of thermophilous species has currently increased. This results mainly from range shift to the northern direction and to higher altitudes.
3. In result of changing climatic conditions, the status of some phytophagous species can change, the role of some species can increase, while the other can decrease.
4. The number and the role of phytophagous species overwintering in egg stage have increased in comparison to species that overwinter in other development stages. It relates to average temperature increase in winter as larvae, pupae and the adults show higher mortality at that time, whereas eggs have probably higher resistance to low temperatures.
5. Stress, which results from water shortage, is one of the global warming consequences. It can have varied influence on phytophagous insects population dynamic. In general, species developing in wood present positive reactions to moderate decrease of humidity. Humidity shortage negatively influences species that suck sap from plant tissues as well as gall-makers. Research regarding typical folio phages as well as leaf-miners does not give unequivocal results.

6. Climate change together with constantly increasing trade can be conducive to invasive phytophagous insect species. Absence of effective natural enemies in new areas and higher plasticity of invasive species in comparison to native species can cause higher level of damage in forest ecosystem.

References

Awmack, C. S., Woodcock, C. M.and Harrington, R. (1997). Climate change may increase vulnerability of aphids to natural enemies. *Ecological Entomology*, **22:** 366–368.

Ayres, M. P.and Lombardero,M.J. (2000). Assessing the consequences of global change for forest disturbance from herbivores and pathogens. *The science of the total environment,* **262(3):** 263–286.

Bale, J. S., Masters,G.J., Hodkinson, I.D, Awmack, C., Bezemer, T.M. and Brown, V.K (2002). Herbivory in global climate change research: direct effects of rising temperature on insect herbivores. *Glob. Change Biol.* **8,** 1–16.

Bale, J. S., Masters,G. J., Hodkinson,I. D., Awmack,C., Bezemer,T. M., Brown Butterfield,J., Buse,A., Coulson,J.C., Farrar,J., Good,J. E. G., Harrington,R., Hartley,S., Jones,T.H., Lindroth,R.L., Press,M. C., Symioudis, I., Waltt,A. D.,Whittaker, J. B. (2002). Herbivory in global climate change research: direct effects of rising temperature on insect herbivores. *Global Change Biology*, **8(1):** 1–16.

Battisti, A. (2008). Forests and climate change – lessons from insects. *iForest,* **1**: 1–5. http://www.sisef.it/iforest/pdf/Battisti_210.pdf [20.10.2010].

Battisti, A., Stastny,M., Buffo, E. and Larsson, S.(2006). A rapid altitudinal range expansion in the pine processionary moth produced by the (2003) climatic anomaly. *Global Change Biology,* **12 (4):** 662–671.

Battisti A., Stastny,M., Netherer,S., Robinet,C., Schopf,A.,Roques, A.and Larsson, S.(2005). Expansion of Geographic Range in the Pine Processionary Moth Caused by Increased Winter Temperatures. *Ecological Applications*, **15:** 2084–2096.

Buffo, E., Battisti,A.,Stastny, M. And Larsson, S. (2007). Temperature as a predictor of survival of the pine processionary moth in the Italian Alps. *Agricultural and Forest Entomology*, **9**: 65–72.

Buse, A., Good,J. E. G., Dury,S.,Perrins, C.M.(1998). Effects of elevated temperature and carbon dioxide on the nutritional value of leaves of oak (*Quercus robur L.*) as food for the Winter Moth (*Operophtera brumata L.*). *Functional Ecology*, **12**: 742–749.

CABI's Crop Protection Compendium, http://www. cabicompendium.org/cpc/aclogin. asp?/cpc/economic. asp?. (Retrieved on December 2018).

Capdevila-Argüelles L., Zilletti B. (2008). A perspective on climate change and invasive alien species. Convention on the conservation of European wildlife and natural habitats. http://www.coe.int/t/dg4/cultureheritage/nature/bern/invertebrates/inf05rev_2008_en.pdf [20.10.2010].

Coviella, C. E., and Trumble, J.T.(1999). Effects of Elevated Atmospheric Carbon Dioxide on Insect-Plant Interactions. *Conservation Biology*, **13(4):** 700–712.

Cramer, H.H.(1967). Plant protection and world crop production. Bayer. *Pflanzenschutz-Nachrichten,* **20**: 1–524.

Currano, E. D., Wilf,P., Wing, W.L., Labandeira, C.C., Lovelock, E.C. and D.L. Royer. (2008). Sharply increased insect herbivory during the Paleocene-Eocene Thermal Maximum. *Proceedings of the National Academy of Sciences,* **105 (6):** 1960–1964.

DeLucia, E. H., Casteel,C. L.,Nabity O, P. D. and Neill, B. F. (2008). Insects take a bigger bite out of plants in a warmer, higher carbon dioxide world. *Proceedings of the National Academy of Sciences*, **105**: 1781–1782.

Esper, J., Büntgen,U., Frank, D.C., Nievergelt, D. and Liebhold, A.(2006). 1200 years of regular outbreaks in alpine insects. *Proceedings of the Royal Society, Series B,* **274**: 671–679.

FAO, (2017).http://www.fao.org/faostat/en/#data/EP/visualize Accessed on 10.7.2019.

Hill, J.K., Thomas, C.D. and Blakeley, D.S. (1999). Evolution of flight morphology in a butterfly that has recently expanded its geographic range, *Oecologia*, **121**:165–170.

Hódar, J.A. and Zamora, R.(2004). Herbivory and climatic warming: a Mediterranean outbreaking caterpillar attacks a relict, boreal pine species. *Biodiversity and Conservation,* **13**: 493–500.

Huberthy, A. F., Denno R. F. (2004). Plant water stress and its consequences for herbivorous insects: a new synthesis. *Ecology,* **85(5):** 1383–1398.

Jepsen J, U., Hage,S. B., Ims,R. A.,Yoccoz, N. G. (2008). Climate change and outbreaks of the geometrids Operophtera brumata and Epirritia autumnata in subarctic birch forest: evidence of a recent outbreak range expansion. *Journal of Animal Ecology*, **77(2):** 257–264.

Jönsson A. M., Harding,S., Bärring,L.,Ravn, H. P.(2007). Impact of climate change on the population dynamics of Ips typographus in southern Sweden. *Agricultural and Forest Meteorology*, **146:** 70–81.

Kalaiselvi,V. (2012). An Economic Analysis of Crop Diversification in Tamilnadu. *International Journal of Current Research and Review,* **4**(8): 147–153.

Karolewski P., Grzebyta,J., Oleksyn, J. andGiertych, M. J. (2007). Effects of temperature on larval survival rate and duration of development of Lymantria monacha (L.) on needles of Pinus silvestris (L.) and of L. dispar (L.) on leaves of *Quercus robur* (L.). *Polish Journal of Ecology*, **55(3):** 595–600.

Logan J. A., Régnière, J. and Powell, J.A.(2003). Assessing the impacts of global warming on forest pest dynamics. *Frontiers in Ecology and the Environment*, **1** (3): 130–137.

Menendez, R. (2007). How are insects responding to global warming. Tijdschrift voor *Entomologie*, **150:** 355–365.

Menéndez, R., González-Megías, A. G., Collingham, Y., Fox, R., Roy, D. B., Ohlemüller, R. and Thomas, C. D. (2007). Direct and indirect effects of climate and habitat factors on butterfly diversity. – *Ecology*. **88:** 605–611.

Moore, B.A. and Allard, G. B. (2008). Climate change impacts on forest health. Forest Health & Biosecurity Working Papers FBS/34E. Forest Resources Development Service, Forest Management Division, FAO, Rome.

Netherer,S and Schopf, A.(2010).Potential effects of climate change on insect herbivores general aspects and a specific example (Pine processionary moth, Thaumetopoea pityocampa). *Forest Ecol Manag.* **259(4):**831–838.

Oerke, E.C. (2006). Crop losses to pests. *Journal of Agricultural Science*, **144:** 31–43.

Oerke, E.C., Dehne,H.W., Scho¨nbeck, F. andWeber, A. (1994). Crop Production and Crop Protection – Estimated Losses in Major Food and Cash Crops. Amsterdam: Elsevier Science.

Palanisami, K., Paramasivam,P., Ranganathan, C.R., Aggarwal,P.K. and Senthilnathan,S.(2009). "Quantifying Vulnerability and Impact of Climate Change on Production of Major Crops in Tamil Nadu, India", From Headwaters to the Ocean-Hydrological changes and Watershed Management, Taylor & Francis Group, London, U.K, ISBN 978-0-415-47279-1 pp 509-514.

Parmesan, C. (1996). Climate change and species' range. *Nature,***382**: 765–766.

Parmesan, C. and Yohe, G.(2003). A globally coherent fingerprint of climate change impacts across natural systems. *Nature,***421**: 37–42.

Petzoldt, C. and Seaman. (2006). Climate change effects on insects and pathogens. Climate Change and Agriculture: Promoting Practical and Profitable Responses, III: 6–16. http://www.climateandfarming.org/clr-cc.php [22.10.2010].

Ramchandra Yadav and Niann-Tai Chang. (2014). Effects of temperature on the development and population growth of the melon thrips, *Thrips palmi*, on eggplant, *Solanum melongena Journal of Insect Science*, **14(1):** 78–81.

Regniere, J. (2009). Predicting insect continental distributions from species physiology. Unasylva, **60(1/2):** 37–42.

Rouault, G., Candau,J.N., Lieutier,F., Nageleisen,L.M., Martin,J.C.,Warzée, N.(2006). Effects of drought and heat on forest insect populations in relation to the (2003) drought in Western Europe. *Annals of Forest Science*, **63 (6):** 613–624.

Sefrova, H. (2003). Invasions of Lithocolletinae species in Europe – causes, kinds, limits and ecological impact (Lepidoptera, Gracillariidae). *Ekologia* (*Bratislava*), **22** (2): 132142.

Sharma, H.C. (2014).Climate change Effects on Insects: Implications for Crop Protection and Food Security. *Journal of a crop Improvement*. **28(2):** 229–259.

Solarimpulse foundation, (2018) https://solarimpulse.com/global-Warmingsolutions?gclid).

Szujecki, A. (1998). Entomologia leśna. Warszawa, Wyd. SGGW.

Thomas, F. M.(2008). Recent advances in cause-effect research on oak decline in Europe. CAB Reviews: Perspectives in Agriculture, Veterinary Science, Nutrition and Natural Resources, 3, 037. http://www.cababstractsplus.org/cabreviews [24.06.2013].

Vanhanen, H., Veteli,T. O., Päivinen,S., Kellomäki, S. and Niemelä, P. (2007). Climate Change and Range Shifts in Two Insect Defoliators: Gypsy Moth and Nun Moth – a Model Study. *Silva Fennica*, **41 (4)**: 621–638.

Virtanen, T., Neuvonen,S., Nikula,A., Varama, M. and Niemelä, P. (1996). Climate Change and the Risks of Neodiprion sertifer Outbreaks on Scots Pine. *Silva Fennica*, **30 (2–3):** 169–177.

Walther, G.R., Post,E., Convey,P., Menzel,A., Parmesan, C., Beebee, T. J. C., Fromentis, J.M.,Hoegh-Guldberg, O. and Bairlein,F. (2002). Ecological responses to recent climate change. *Nature*, **416:** 389–395.

Williams, D. W. and Liebhold, A. M.(1995). Forest defoliators and climatic change: potential changes in spatial distribution of outbreaks of western spruce budworm (Lepidoptera: Tortricidae) and gypsy moth (Lep: Lymantriidae). Environmental entomology, 24 (1): 1–9. Williams D. W., Liebhold A. M. 1995b. Herbivorous insects and global change: potential changes in the spatial distribution of forest defoliator outbreaks. *Journal of Biogeography*, **22 (4/5):** 665–671.

13

Coastal Biodiversity of India

K. Kathiresan and S. Mohan
Centre of Advanced Study in Marine Biology
Annamalai University, Parangipettai
Department of Civil Engineering, Faculty of Engineering and Technology
Annamalai University, Annamalai Nagar

Coastal Ecosystems of India

India has a long coastline of 7,517 km that span 13 maritime states and union territories, surrounded by the Indian Ocean, Arabian Sea and Bay of Bengal. It has an Exclusive Economic Zone of 2.02 million square kilometers. The country has a diverse range of habitats: continental area and offshore islands and a variety of coastal ecosystems such as estuaries, lagoons, coral reefs, mangroves, salt marshes, sea grass beds, backwaters, rocky coasts, mudflats, sand dunes, and coastal beaches. Coasts and islands form 2.8% of India's geographical area. The coastal ecosystem consists of 43,230 km^2 of coastal wetlands with 97 major estuaries and 34 major lagoons; 4921 km^2 of mangroves with 31 mangrove areas, and 3062.97 km^2 of coral reefs with 5 coral reef areas (MoEFCC, 2018). A network of 14 major, 44 medium and numerous minor rivers together with their tributaries with a total length of over 40,000 km, contribute to the coastal sea directly or indirectly. By area, tidal/mudflats are predominant and by length, more than half of the Indian coastline is sandy. The states of Gujarat, Tamil Nadu and Andhra Pradesh have extensive coastline (Fig. 1).

There is a remarkable difference between east and west coasts of India. The east coast is bestowed with beaches, lagoons, deltas and marshes, whereas the west coast is exposed to heavy surf and rocky shores. The east coast experiences with only a weak upwelling associated with the northeast monsoon (Oct-Jan), while the west coast is the region of intense upwelling associated with southwest monsoon (May-Sep.). All the islands on the east coast are continental islands, whereas the major island formations in the west coast are oceanic atolls (Venkataraman and Wafar, 2005). The general current patterns of seas around India follow a general clockwise circulation during the southwest monsoon and counter clockwise circulation during the Northeast monsoon. Salinity also differs: in the Arabian Sea, highly saline water masses due to high ratio of evaporation and precipitation, whereas the Bay of Bengal has comparatively low salinity due to high runoff and precipitation (Venkataraman, 2007).

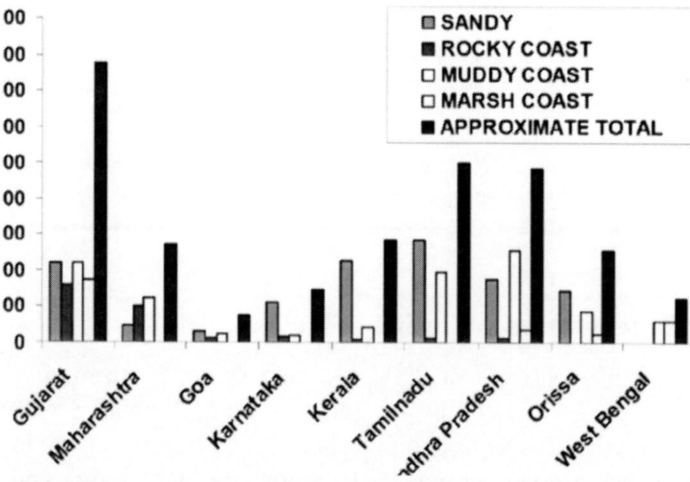

Figure 1. Length (km) of major ecosystems in coastal states of India (*Space Applicaton Center, 1992; Venkataraman and Wafar, 2005*)

Uniqueness of Coastal Biodiversity

India is one among 18 mega-biodiversity countries and 25 hotspots of the richest eco-regions of the world. Among the Asian countries, India is the only one that has a long record of inventories of coastal and marine biodiversity dating back to at least two centuries (Venkataraman and Wafar, 2005). The country is rich in marine and coastal biodiversity, and this is due to habitat diversity which includes almost all types of intertidal habitats, from hypersaline and brackish lagoons, estuaries, and coastal marsh and mud flats, to sandy and rocky shores. A number of factors including the long coast line, tropical climate, and nutrients supplied by rivers along the coast, have combined to produce a variety of biologically rich and productive coastal and off-shore marine ecosystems. The most spectacular natural treasures are (i) Inter-tidal mudflats teeming with migratory birds, (ii) dense mangrove forests inhabited by the endangered tiger, (iii) delicate sea grass meadows favoured by the seacow (dugong), (iv) most beautiful corals colonized with ornamental fishes, (v) sandy coast with the world's largest nesting site of sea turtles, and (vi) rough sea migrated with the largest whale shark fish (Kathiresan and Qasim, 2005).

India is unique to have globally threatened species such as Royal Bengal tiger, sea turtles, fishing cat, estuarine crocodile, the Gangetic dolphin, and river terrapin. The mangrove ecosystem has a total of 4,017 biological species that include 925 flora and 3,182 fauna, perhaps the largest record in the world (Kathiresan, 2000, Kathiresan and Qasim, 2005; Kathiresan, 2018). There are 218coral species representing all the three major reef types (atoll, fringing and barrier reefs), 14 species of seagrasses and over 1000 species of seaweeds. There are nearly 2 million water birds of about 200 species migrate in winter (Kathiresan, 2000; Venkataraman and Wafar, 2005). Five of the world's seven species of sea turtles are found in India, namely the Olive Ridley (*Lepidochelys olivacea*), Green (*Chelonia mydas*), Hawksbill (*Eretmochelys imbricata*), Loggerhead (*Caretta caretta*) and the critically endangered Leatherback (*Dermochelys coriacea*) turtles (Rajagopalan, 1996). Odisha on the east coast of India has the world's largest mass nesting Olive Ridley Turtles during the months of October to April, supporting a nesting population of more than half a million individuals of the species. The threatened whale shark (*Rhincodon typus*), the largest fish species in the world can be found off the coast of Saurashtra in Gujarat, while the last population of one of five threatened sub-species of Asiatic wild ass (*Equus hemionuskhur*) occurs in the salt marshes of the Little Rann of Katchchh in Gujarat (Kathiresan and Qasim, 2005). The overall area cover of coral reefs in India is 3,062.97 sq km. This includes lagoons of 521.5 sq. km and coralline shelf of 157.6 sq. km. Fringing reefs are present in Gulf of Mannar and Palk Bay, platform reefs are seen along the Gulf of Kachchh, while atoll reefs occur in the Lakshadweep archipelago. Fringing and barrier reefs are present in Andaman and Nicobar islands. Coral patches are present along the west coast of India, in Kerala, Karnataka, Goa and Maharashtra (MoEFCC, 2018).

Importance of Coastal Biodiversity

The coastal ecosystems are extremely important assets, as they provide a wide range of ecosystem goods and services. Coastal areas of India are densely populated, and about 30% of its human population is dependent on coastal and marine resources. Over 250 million people live within a distance of 50 km from the coast, and a large proportion of them are in urban centers such as Mumbai, Chennai and Kolkata. The population density is also much more in coastal areas than the national average. For example, in the state of Tamil Nadu, the population density in coastal areas is 528 per square km against the state average of 372 per square km. In Mumbai, Kolkata and Chennai, the population density ranges from 20,000 to 50,000 per square km. The people living in the coastal belt rely on resources for their livelihood through fishing and other economic activities that include agriculture, tourism, ports, maritime shipping, and communication sectors and their related infrastructures. India is a major exporter of seafood. The fisheries sector provides employment to more than six million people and it accounts for 1.2% of India's total GDP and 5.4% of total agricultural GDP (Venkataraman, 2007). Marine Fishing is a major livelihood for about 4.3 million people belonging to 18,340 families in 8,64,550 marine fishermen households of 3,288 marine fishing villages, distributed among the nine maritime states and two union territories. Of these, 90% belong to traditional fisher families. Tamil Nadu has the largest population of marine fishers (0.8 million) followed by West Bengal (0.634 million and Kerala (0.61 million) (MoEFCC, 2018). The importance of maintaining healthy coastal and marine ecosystems for coastal protection against natural disasters as well as for general human well-being was reaffairmed after the 1999 Super cyclone of Orissa and the 2004 tsunami of Indian Ocean region.

Climate Change and Vulnerability

Most of the Indian coastline is safe, but certain coastal areas are vulnerable at locations such as Mumbai (Maharashtra), Alappuzha (Kerala) along the west coast, and Kendrapada (Odisha) and Sagar Island (West Bengal) on the east coast. Islands like Sagar Island are twice as vulnerable to hazards. The east coast is more vulnerable than the west coast due to starvation of water and sediments from large river systems and increased intensity of coastal flooding and erosion. The most vulnerable villages due to extreme climate events are Kendrapada, Nellore and Nagapattinam in east coast. Due to changes in the coastline and sea level fluctuations, several ancient ports and settlements have either submerged in the sea like Bet Dwarka, Pindara, Poompuhar and Mahabalipuram, or pushed far in the hinterland such as Harappan sites along the Gujarat coast (MoEFCC, 2018).

Water resources in Indian coast are vulnerable to climate change. Floods, rising sea levels and other coastal hazards contaminate freshwater sources, damaging drinking water supply, wastewater collection and treatment systems. In coastal areas, when a large quantity of freshwater is removed from rivers and aquifers, saltwater will move upstream into the river mouth and the aquifers. All these challenges put high pressure on the availability of freshwater on the coast (MoEFCC, 2018).

Estuaries are generally the source for atmospheric CO_2 due to dominant heterotrophic condition, whereas shelf regions are the sink for the atmospheric CO_2 due to dominant auto trophic condition. However, several local processes such as coastal upwelling, river discharge and atmospheric deposition of pollutants can also change the pattern of CO_2 source/sink in the coastal regions.

The annual emission of CO_2 from Indian estuaries and shelf regions is estimated at about 2 TgC and 6.33 TgC to the atmosphere respectively. Fertilization using natural nutrients may be attempted to increase primary production of the estuaries and shelf regions for enhanced removal of atmospheric CO_2. This must be done with care of not forming oxygen minimum zones. In addition to estuaries and shelf regions, coastal ecosystems also emit CO_2 to certain extent. For example, the net emission is at 5106 GgCO_2e/yr from mangroves and surrounding waters, and 41 Gg CO_2e/yr from sea grass ecosystems of India. Mangroves along the east coast contribute to 2735 GgCO_2e/yr than mangroves along the west coast (1631 GgCO_2e/yr). The anthropogenic influences and seasonality intensifies heterotrophic mineralization in coastal ecosystems, leading to increased green house gas emission. Therefore, there is a need for conservation and restoration of the blue carbon ecosystems to enhance the carbon sequestration potential and to contribute to the climate change mitigation efforts. However, the coastal ecosystems are three times more efficient in capturing atmospheric CO_2 than the terrestrial forest ecosystems. For example, the mean C stock in top one meter mangrove soil varies from 62 to 207 MgC/ha whereas, C sequestration in sea grass sediment ranges between 107 and 143 MgC/ha (MoEFCC, 2018).

Sandy beaches are highly vulnerable to changes due to storm surges, which in turn affect the marine turtle nesting sites and other intertidal faunal dependent on the beaches. Among all the turtle nesting sites along the coast of India, North of Bitikola Nadi in Odisha experiences an average erosion of -449.67 m and the net shoreline movement (NSM) at a rate of -17.33 m/yr. Regarding mangroves, the coastal erosion is observed maximum in Andhra Pradesh specifically in Krishna Delta with an average erosion of -203.6 m and the NSM at -9 m/yr followed by Sundarbans (-92.6 m and NSM at -5.1 m/yr) in West Bengal. Some of the turtle nesting sites in India are highly vulnerable due to erosion. A few of the dense mangrove ecosystems along Krishna-Godavari Deltas are lost due to severe coastal erosion (MoEFCC, 2018).In India, major reef areas such as Gulf of Mannar and Palk Bay, Gulf of Kachchh, Andaman and Nicobar Islands and Lakshadweep Islands have experienced coral bleaching events and consequent mortalities, since 1998 during the warmest years. Coral degraded reef areas should be restored with resilient and resistant native coral species to assist the recovery process. However, the commercially important oil sardine *Sardinell longiceps* has extended its northern and eastern boundaries of distribution, resulting in expanded fishery all along the Indian coast within the last three decades. Prime cause of this is seawater warming (MoEFCC, 2018).

Need for increased conservation

India is strong in policy and legal frame work in conservation of marine and coastal biodiversity. India has designated 25 marine protected areas (MPAs) in Peninsular India covering 8,231 km^2 and 106 MPAs in the Andaman and Nicobar and Lakshadweep Islands. Several Acts laws and notifications have provided a protection to the coast of India including its island territories. The Wildlife Protection Act (1972) provides for protection of marine species and coral reefs. Coastal Regulation Zone Notification (2011) under the Environment Protection Act (1986) regulates onshore development activities, which affect coastal environments. A coupled management regulatory approach such as the coastal zone management plans (CZMP) and integrated coastal zone management (ICZM) is available for better conservation and management of coastal India, focusing on three aspects namely environmental sustainability, economic viability and social acceptability. This also helps to provide a protection against climate change induced pressures at the land-sea interface (MoEFCC, 2018).

In spite of all the efforts, the increased population pressure has led to resource depletion, and environmental degradation due to pollution caused by disposal of agricultural, domestic and industrial wastes. Overexploitation of fish species and associated destructive harvesting practices through use of inappropriate fishing gear are important issues. The use of fish traps made of long-lasting materials with small mesh sizes results in the capture of juveniles affecting future populations of fish. Marine fisheries in India is mainly supported by a few commercially important targeted species, which form the bulk of the landings, namely Indian mackerel, anchovies, seer fish, ribbonfish, Bombay duck, carangids, elasmobranches, sciaenids, perches, silver bellies, lizard fish, penaeid shrimps, cephalopods and bivalve molluscs. Along with increase in the targeted catch, a number of untargeted fish and other biota are removed from their habitat and discarded as waste. Shrimp trawlers probably have the highest rate of by-catch bringing in up to 90% or more of "trash fish". Gill nets used to catch fish also bring in a host of other animals like dolphins and turtles. Economic developmental activities have made conversion and destruction of coastal ecosystems for construction of roads, ports, tourist resorts, aquaculture and other forms of land use. Regarding natural threats, cyclonic disturbances are frequently occurring during October-November along the coast. Also freshwater runoff kills many fauna and flora in semi-enclosed bays and lagoons by lowering salinity and depositing large amounts of sediment and nutrients. Additionally, climate changes associated with global warming, sea level rise and intensified cyclones are expected to have negative impacts on coastal and marine ecosystems (Venkataraman and Wafar,

2005; Venkataraman, 2007). After tsunami 2004, the coastal soil salinity has increased, which changes the floral species composition and adversely affects the benthic organisms in the mangrove sediments particularly along the east coast (Sandilyan *et al.*, 2010). Hence, there is a pressuring need for conservation of diversity in the coastal ecosystems.The number of species known in the coastal and marine environment is in the order of over 15,000. The inventory is largely available for commercially important groups of fishes or molluscs and it is extremely poor for minor phyla or microbes. Only two-thirds of the total marine habitat has been covered till today and the remote islands and other minor estuaries still remain unexplored. The inventory is several times higher than what is known today. Lack of trained taxonomists is a serious drawback to achieve this (Venkataraman and Wafar, 2005). Without proper data, it is not possible for conservation and sustainable utilization of coastal biodiversity in the present context of increasing pressure on biological resources.

Acknowledgements

The authorsare thankful to UGC, New Delhi for BSR Faculty fellowship and the authorities of Annamalai University for providing facilities as well Dr. M Prakash, Faculty of Agriculture.

References

Kathiresan, K. (2000). Mangrove atlas and status of species in India. Report submitted to Ministry of Environment and Forest, Govt. of India, New Delhi, 235 pp.

Kathiresan, K. (2018). Mangrove Forests of India. *Curr. Sci.*, **114(5):** 976–981.

Kathiresan, K. and Qasim, S.Z. (2005). In Biodiversity of Mangrove Ecosystems. Hindustan Publishing Corporation, New Delhi, 251 pp.

MoEFCC, (2018). Climate Change and the Vulnerable Indian Coast. (Ed. R. Ramesh and JR. Bhatt), Ministry of Environment, Forest, Climate Change, Govt. of India, New Delhi. pp.391.

Rajagopalan, M. (1996). The marine turtles and their conservation and management. In: Menon N.G and C.S.G. Pillai, (eds.). Marine Biodiversity: Conservation and Management, Central Marine Fisheries Research Institute, Cochin, pp. 126–132.

Sandilyan, S., Thiyagesan, K., Nagarajan, R. and Jayashree Vencatesan. (2010). Salinity rise in Indian mangroves – a looming danger for coastal biodiversity. *Curr. Sci.,* **98:** 754–756.

Venkataraman, K. (2007). Conservation and management of coastal and marine ecosystems in India. National Biodiversity Authority, Chennai, 91 pp.

Venkataraman, K. and Wafar,M.(2005). Coastal and marine biodiversity of India. *Indian J. Mar. Sci.,* **34(1):** 57–75.

14

Coastal Agroforestry: Challenges and Opportunities

Masilamani, P. C. Buvaneswaran* and A. Alagesan
Anbil Dharmalingam Agricultural College and Research Institute, Tamil Nadu Agricultural University, Tiruchirappalli, 620 027, Tamil Nadu.
**Institute of Forest Genetics and Tree Breeding, ICFRE, Coimbatore-641002, Tamil Nadu.*

Introduction

India contains around 7517 km of coast line, which is surrounded by the Arabian Sea on the west, the Bay of Bengal on the east, and the Indian Ocean to its south. The Indian Coastal ecosystem, where land and water join to create an environment with a distinct structure, diversity, and flow of energy. They include salt marshes, mangroves, wetlands, estuaries, and bays and are home to many different types of plants and animals. (Marale. 2013). The geographic and land use details of Indian coastal zone is presented in Table 1.

Table 1: Geographic and Land Use details on Coastal zone in India

Coastal data	
Length of coastline	7516.6 km (Mainland: 5422.6 km; Island Territories: 2094 km)
Total Land Area	3,287,263 km^2
Area of continental shelf	372,424 km^2
Territorial sea (up to 12 nautical miles)	193,834 km^2
Exclusive Economic Zone	2.02 × 106 million km^2
Maritime States and UT	
Number of coastal States and Union Territories	Nine states; Gujarat, Maharashtra, Goa, Karnataka, Kerala, Tamil Nadu, Andhra Pradesh, Odisha, West Bengal Two Union Territories: Daman & Diu, Puducherry
Island Territories	1. Andaman & Nicobar Islands (Bay of Bengal) 2. Lakshadweep Islands (Arabian Sea)
Total number of coastal districts	66 coastal districts in mainland India; 3 in Andaman & Nicobar and 1 in Lakshadweep

Coastal Geomorphology (Mainland)	
Sandy Beach	43 %
Rocky Coast	11%
Muddy Flats	36%
Marshy Coast	10%
Coastline affected by erosion	1624.435 km mainland 132 (islands) (CPDAC)
Coastal Ecosystems	
Coastal wetlands	43230 km^2
Major estuaries	97
Major Lagoons	34
Mangrove Areas	31
Area under mangroves	6740 km^2 (57% East coast,23% west coast, 20% Andaman &Nicobar Islands)
Coral Reef Areas	5
Marine Protected Areas	31
Area Covered by MPA	6271.2 km^2

In coastal systems, poor resource management is among the main causes of its degradation. As such, impacts arising from climate change, including sea-level rise, has forced an increase in the demand for sustainable coastal ecosystem science to inform management decisions. The realization of current and future sustainability objectives depends on the development and implementation of coherent strategies on managing dynamic ecosystems for retaining their ability to undergo disturbance, while maintaining their services, functions and control mechanisms (Ali *et al.*, 2018). This paper provides insight into the various Agroforestry strategies that are suitable for coastal areas and also discusses the challenges and opportunities for adaption.

Agroforestry

Agroforestry is an integrated approach for maximization of agricultural production per unit of land. This has been practiced in different forms since ancient times but research work on evaluation of scientific principles supporting this land use has been overlooked. However, recent years are witnessing a rejuvenated interest in agroforestry among researchers, developmental experts and policy planners. The pertinent reasons for coming into limelight again are its uniqueness in sustaining productivity and fertility of soil and its role in protecting environment against degradation by intensive agriculture. There are many traditional Agroforestry practices involving combined production of trees and agricultural species on the same piece of land. However, on intensification of agriculture, particularly in India, whereby capital penetrates agricultural production in order to obtain maximum profit, trees in farmlands had been looked as an interfering component and removed from the agriculture land use systems. However, agroforestry systems potentially provide options for improvement in livelihoods through simultaneous production of food, fodder, timber and firewood as well as has potential in meeting India's commitment in Paris Climate Agreement, 2015 towards fulfilling nationally determined contribution earmarked for forest sector. Further, Agroforestry also can play a role in increasing farmer's income while simultaneously meeting the huge demand for wood and in turn minimizing the India's annual imports of logs and wood products – which have increased from \$500 million to \$2.7 billion (Indian ₹30,000 crore) over the past decade.

Recognizing the potential of agroforestry as a land use for economic upliftment of farmers, sustainable agriculture production with resilience to climate change, the Government of India launched the National Agroforestry Policy in 2014. The major strategies proposed in the Policy to achieve these goals are:

(a) Establishment of institutional setup at national level to promote agroforestry,
(b) Simple regulatory mechanism for harvest and sale of wood from agroforestry,
(c) Improving farmer's access to quality planting materials,
(d) Providing institutional credit and insurance cover for agroforestry,
(e) Strengthening farmers' access to markets for tree products. Further, at present agroforestry meets almost half of the demand of fuel wood, 2/3 of the small timber, 70-80 per cent wood for plywood, 60 per cent raw material for paper pulp and 9-11 per cent of the green fodder requirement of livestock, besides meeting the subsistence needs of households for food, fruit, fiber, medicine etc (Dhyani, 2018).

With regard to Agroforestry systems for coastal zone is mainly aimed to address three major problems being faced by farmers in this region. These three issues are (i) strong winds with high speed leading to crop damages, (ii) Soil salinity and (iii) Water logging due to poor drainage and shallow water table in this region. To address these three key problems of the coastal zone, the potential agroforestry systems identified are:

I – Windbreaks and Shelter belts: to minimize crop damage due to wind

II – Planting of salt tolerant tree species in reclaiming salt affected soils

III – Biodrainage systems to address the problem of water logging.

This paper also briefs on other potential agroforestry systems for the coastal zone with suitable tree species, crops and management practices.

I – Windbreaks and Shelterbelts for Coastal zone

With regard to potential of agroforestry systems for the coastal zone, shelter belt and Windbreaks are prominent agroforestry system to be promoted for the multiple benefits like i) reduced wind speed, ii) reduced evapo-transpiration, iii) reduction in soil erosion, iv) increased crop productivity and v) wood supply. A windbreak is a narrow row of trees planted in fields bordering a farm plot (IDF, 1981). Webster on-line dictionary defines windbreak as 'Hedge or fence of trees designed to lessen the force of the wind and reduce erosion'. It is reported that the term "Windbreak" was first used in popular English literature sometime before 1886. It is more popular and more widely used in developed countries like USA, UK, USSR, Germany than in developing countries like India. On the contrary, windbreak as an Agroforestry system has greater and more vital role to play in developing countries, more particularly with reference to increasing land area under Agroforestry with minimal sink in cultivation area under agriculture

A shelter belt is defined as belt of trees and/or shrubs maintained for the purpose of shelter from wind, sun, snowdrift etc. Shelter belts are generally more extensive than windbreaks and cover areas larger than a single farm and sometimes a whole region on a planned pattern (Dwivedi, 1992). However, 'Windbreak' has been used as a broad term to include both shelterbelts as well as any other vegetative environmental buffers (Tyndall and Wallace, 2011). 'Windbreaks' being broader term and is commonly used in the context of Agroforestry system, this chapter uses only 'Windbreaks' as a common term for both windbreaks and shelter belts.

Benefits of Windbreaks reported in Research studies

Windbreaks on field boundaries effectively control injuries to the tender crops from sand blasting and hot wind. It reduces wind velocity by 20-46% and soil loss by 76% (Gupta *et al.* 1997). Windbreaks are an important tool for farming in semi-arid areas. In the western Rajasthan, 3-row windbreak (of *Cassia siamea, Albizia lebbek* and *C. siamea*) is highly effective in reducing wind speed and loss of nutrients. Hodges *et al.* (2004) found the wind protection provided by shelterbelts (tree windbreaks) can increase pod yields of *Phaseolus vulgaris* both early and late in the seasons. In Egypt, it has been estimated that sheltered field of cotton, wheat, summer maize and rice increased yield by 36, 38, 47 and 10% respectively, over the unprotected fields. Besides crop protection and enhanced productivity, the windbreaks contribute to the organic matter content of the soil through leaf fall (Onuegbu, 2002).

There are several outstanding examples of successful windbreaks around the world. Since 1949, a 3000 km long windbreak of *Casuarina equisetifolia* has been established in China along the coast bordering the South China Sea. This has provided shelter for crops growing on the leeward side and has also stabilized drifting sand in the area (Turnbull, 1983). In Egypt, Casuarina shelter belts have also been grown to protect agricultural land mainly from wind erosion (El-Lakany, 1981).

Campi *et al.* (2009) studied the effects of tree windbreak on microclimate and wheat productivity in a Mediterranean environment. The 3-year field study was carried out by integrating agronomic and microclimatic approaches. Further, within the protected area, wheat water use efficiency (WUE, calculated as the ratio between yield and seasonal evapotranspiration) attained the maximum value of 1.15; out of the windbreak protection, WUE was 0.70 kg m^{-3}. Since windbreaks reduce ET, farms of the Mediterranean environments should be re-designed in order to consider the windbreaks as possible issue of sustainability.

Ideotype Breeding for Windbreaks: Suitable ideotype for 'Windbreak and Shelter belt' agroforestry systems need to have more branches and foliage biomass and should not have self pruning ability. The branch angle also needs to be near right angle. During 2014, Institute of Forest Genetics and Tree Breeding (IFGTB), Coimbatore made such effort in selection of tree varieties for windbreaks and has released five superior clones of *Casuarina junghuniana* most suited for providing protection against windstorms to horticultural and agricultural crops like banana, citrus crops, red gram, etc. The Names of the clones released are IFGTB-WBC-1, IFGTB-WBC-2, IFGTB-WBC-3, IFGTB-WBC-4, and IFGTB-WBC-5 (Buvaneswaran *et al.* 2018). Windbreak tree varieties of IFGTB show high level of branch persistence with 40 to 50 thick and horizontal branches within 3 m height from the base of the tree. The other superiority of these clones are: i) greater branch thickness (having thickening of branches at the rate of 8 to 11 mm per year), ii) wider branch angle (66 to 82 degrees - near right angle) along with iii) greater height growth rate (2.5 to 3 m per year) and iv) faster diametrical growth rate of main stem (1.5 to 2 cm per year). This Casuarina based Windbreak tree varieties planted on the periphery of farm lands can help slow down the speed of wind and minimize the damage to agriculture crops. This clonal windbreak also reduce evaporation from the soil and also reduce water loss through transpiration from the crops inside the windbreak. This in turn increase productivity of agriculture crops from 10 to 30%.

Windbreaks to increase Agricultural crop productivity

In farm field in Coimbatore, red gram was planted (variety CO - 8) both inside the windbreaks and outside the windbreaks. Yield study was conducted in the field with windbreaks and red gram planted. Observation revealed that this Red gram variety (CO -8) was tall more than a meter and hence lodged due to heavy wind in the open field. However, the crop lodging was prevented inside the windbreaks. Further, the yield of red gram was 1.5 times of that observed outside the windbreaks. Another observation made in this field was that after the irrigation done for the Red gram, the soil moisture was retained for more days in field with windbreaks, when compared to the open field. The increased soil moisture retention is mainly due to reduced water loss inside the field with windbreaks through minimized evaporation from the soil and through less transpiration rate from crop.

Yield of Red gram (kg/acre)	
Field with Windbreak	Open field
600	400

Windbreak with clones of IFGTB in field with red gram (Var CO-8) in Coimbatore district.

In continuation to this efforts in promotion of windbreak, the future research focuses can be on (i) Developing superior phenotypes of species other than Casuarina and also indigenous species for Windbreaks through ideo type breeding programme, (ii) Researches on clone specific Tree-crop interactions under Windbreak Agroforestry systems, (iii) Assessing the potential of Windbreaks for increasing water productivity with particular reference to dry land agroforestry and (iv) Scope of Deploying Windbreaks for increasing fruit-set in fruit orchards and tree seed orchards.Review on choice of species suitable for windbreaks other than Casuarina reveals that there are tree species which are very commonly known to our country has been used as windbreaks viz., (*i*) *Azadirachta indica,* (*ii*) *Thespesia populnea,* (*iii*) *Calophyllum inophyllum* (*iv*) *Erythrina variegate,* (*v*) *Bambusa sp.* (*vi*) *Gliricidia sepium* (*vii*) *Araucaria heterophylla* (*viii*) *Bauhinia racemosa.*

Shelterbelts and shore protection

Natural disasters like tsunamis, storms, cyclones, coastal erosion, typhoons, sea level etc., causes damages to coastline of the world. The above natural phenomena devastate the human lives, livestock, trees & forests demolish constructions, welfare structures, Loss of agriculture land, aquaculture, saltpan, coastal land, beaches, tidal land etc. Since cyclones and tsunamis mostly originate from the sea and move towards the land , the wind – generated reduced , reflected , deflected and dissipated, when they pass through obstacles such as coastal forest, mangroves, offshore islands, coral reefs, head lands, sea cliffs, sand pits, mud flats, sand dunes, creeks etc. Other than the above natural features, manmade structures along the coastal line such as offshore platforms, sea wall constructions, harbour and plantations also help in energy dissipation.

Types of Shelterbelt

(a) Hard Defenses: Static shoreline structures such as those constructed from timber, steel, concrete, asphalt and rubble. These involve linear structures such as sea walls, revetments and control structures of artificial headlands, offshore breastwork and groynes.

(b) Soft Defenses: Mobile/responsive defense measures, seek to work with nature rather than control it. Such structures may consist of sand or shingle beaches and dunes or banks, which may be natural or constructed, and may include control structures. These can include soft solutions of beach nourishment, cliff/dune stabilization, bypassing and managed retreat.

II - Agroforestry for salt affected soils in coastal zone

Another problem in coastal zone is salt affected soils and physio-chemical properties of such salt-affected soils normally reflect the amount and type of salts present. Principal criteria used to identify the dominant constraints relating to the salt affected soils are (i) salinity of the saturation extract determined from ECe at 25°C, (ii) exchangeable sodium percentage (ESP), and (iii) pH of the saturated soil paste. Salt-affected soils are thus, broadly classified into saline and alkali soils (Abrol and Sandhu,1980; Szaboles, 1980). The basic problem of saline soil given in Table 2.

Table 2: Properties of saline soil

Properties	Saline soils
ECe at 25 °C dS m^{-1}	4.0 or more
pH, saturated soil paste	Less than 8.2
ESP	Variable

Properties	Saline soils
Chemistry of soil solution	Dominated by chlorines and sulphates of sodium, calcium, magnesium
Effect of electrolytes on soil particles	Aggregation
Main adverse particles on plants	High osmotic pressure of solution
Geographical distribution	Associated mainly with arid and semi arid areas
First aim of reclamation	Removal of excess electrolytes through leaching

Establishments of tree plantations

About 40% of the barren salt affected soils of India are confined to the Indo – Gangetic plains. High alkali status of the soils are reported to be the major constraints impairing their productivity (Abrol and Bhumbla, 1978). Most of these soils have been formed under the influence of sodium carbonate. Hydrolysis of this salt imparts high pH and ESP which affect the physio-chemical and biological properties of such soils adversely. Presence of calcic horizon of about 40-60 cm wide between deep rooted plants. Planting of trees on such soils would, thus help maintaining healthy agro-ecological system in the region known for its higher food production capacity.

Considering the vast scope of these soils for afforestation, available research information is inadequate. A review of research on afforestation of salt affected soil indicates that the few tree species can be grown successfully by adopting appropriate management practices. But lack of systematic experimental evidences regarding adaptability and tolerance of different species to varying degree of presence of alkali in soil hampering planning and planting operations. In several of such investigations in India and abroad, tree species were reported (Yadav, 1980) to differ widely for their tolerance to alkali and saline soils. But actual data relating to soils on which tree species were grown were mostly unavailable. Consequently, evaluation of critical tolerance limits was not feasible. Moreover, suitability of forest species is also to be decided in respect of local agro climate, purpose and nature of afforestation operations and other soil management factors.

To overcome the physio-chemical constraints, which makes the salt affected soils inhospitable for economic growth of most of the tree species, an appropriate modification in planting techniques suited to normal soils is required for achieving reasonably high survival percentage and rapid early grown of trees in salt affected soils. The given modification should ensure (i) availability of soil environment for optimum root growth, (ii) manipulation for leaching of soluble salts from the root zone, (iii) maximum retention of available soil moisture within root zone, (iv) perforation of any type of hard pan or mechanical impediment present in subsoil and (v) maintenance of appropriate level of soil fertility through fertilizers and organic manures. In view of these considerations, planting of trees for establishing agroforestry systems on salt affected soils requires a correct choice of species in addition to any special treatment for the site preparation.

Promising tree species for saline soils

The success of the vegetation programmes in saline soils depends on kind, content and distribution of salts in soil profiles and the prevailing moisture regimes particularly during the early critical growth period of tree saplings (Table. 3)

Table 3. Tree species tolerant to soil salinity

Tolerant	Moderate salinity	sensitive
Prosopis chilensis	*Eucalyptus umbellata*	*Leucaena latisiliqua*
Tamarix troupii	*Terminalia arjuna*	*Morus species*
Tamarix aphylla	*Pongamia pinnata*	*Salix babylonica*
Acacia tortillis	*Dalbergia sissoo*	*Populous species*
Acacia nilotica	*Acacia catechu*	*Tamarindus indica*
Casuarina obesa	*Acacia Senegal*	*Melia azedarach*
Casuarina equisetifolia	*Prosopis cineraria*	*Azadirachta indica*
Callistemon lanceolatus		*Albezia lebbek*

(Gill *et al*., 1990)

III - Biodrainage system for water logging in coastal zone

The third prominent issue in the coastal zone is Water logging or poor drainage. Shallow ground water tables and associated salinity problems have become dominant features in agricultural areas around the world. These problems have been caused by increasing pressures on land resources caused by rising populations, especially in irrigation areas. Conventional physical drainage works require expensive capital investment, operation and maintenance. Physical drainage measures also generate drainage effluent. Disposal to rivers of the often saline, and sometimes chemically contaminate deffluent is increasingly considered unacceptable because down stream users in the catch mentrely on these river systems for their water supplies. Any positive alternative, preferably cheaper, addition to our arsenal of drainage techniques would be extremely welcome in our fight to keep ground water tables in our agricultural areas under control. Biodrainage, i.e., the use of vegetation to manage water fluxes in the landscape, is one such technique that has recently attracted interest in drainage and environmental management circles.

Biodrainage relies on vegetation, rather than engineering mechanism store move excess soil water through evapotranspiration. It is often considered attractive because it requires only an initial investment in site development (planting of a "biodrainagecrop") and (potentially) returns a benefit when the biocrop is harvested for fodder, wood or fibre. In addition, under some management scenarios, viz. certain cropping systems and slightly saline conditions, it might offer limited scope to achieve nutrient and/or salt-balance through removal of biomass, thus alleviating the problem of the disposal of polluted drainage effluent from the biodrainage crop area by reducing volumes and improving the quality of the effluent.

Biodrainage for degraded drainage problem lands in Tamil Nadu – A case study

Water logging and soil salinity are of great concern all over the world and in India as well. They pose a major limitation to agricultural productivity. Current estimate shows that about 60 per cent of the land in India suffers from soil erosion, water logging and salinity. Approximately, 7.0 million ha is affected by salinity and alkalinity due to water logging in different agro-climatic regions of the country of which about 54,000 ha. is in the Cauvery delta. The major reasons for this land drainage problem are low discharge capacity of drainage, rivers, flat longitudinal slope (1 in 2000), silting of bed and channels, meandering of drainage courses, presence of aquatic weed growth in drainage channels and absence of link drains. Management strategies like provision of drainage are to be developed to address the above problems.

Drainage is removal of excess water from the soil surface and the soil profile. Generally, engineering approaches, based on the laws of potential flow, such as deep open ditches, vertical drainage or horizontal sub-surface drainage, are adopted. These approaches require expensive capital investment, operation and maintenance. This will also generate drainage effluent, mainly saline in nature, the disposal of which creates environment problems. A possible alternative is the biological drainage or bio-drainage. Bio drainage relies on vegetation, rather than engineering mechanisms to remove excess soil water. This happens through evapotranspiration, which is commonly expressed as "letting the vegetation drink itself out of the water logging problem" (Fig: 1). The system of bio- drainage is low cost and does not require installation of any physical structures. However, an initial investment for planting the vegetation is required. When established, the vegetation will also yield marketable products such as fodder, fibre and wood. To understand the effect of bio-drainage in solving the drainage problems caused by canal seepage, a pilot study was undertaken by Masilamani *et al.*, 2003 at Agricultural Engineering College &Research Institute farm at Tiruchirappalli, Tamil Nadu.

A stretch of a canal of 0.63 km length and 3.0 m width nearing the tank was considered for observation. In addition to the seepage water from the tank, the sewage from the nearby residential area was also discharged into this canal. There was lot of weed growth (Typha sp.) which reduced the free flow of the water disabling its proper use. The stagnation of water resulted in foul smell, increased mosquito population, and increased soil moisture, thus leading to cultivation problems. The discharge of this water into the Pullambadi canal (Irrigation canal of the Cauvery system) added to environmental problems. Hence, bio-drainage system was introduced to solve the drainage problem and to use the water efficiently. Half of the canal length was planted with Eucalyptus tereticornis (Plate.1) at a spacing of 1 m and the other half with different varieties of banana (Poovan, Karpuravalli, Rasthali and Monthan) at a spacing of 3.0 m. Both the cropscan be cultivated continuously by ratooning/ coppicing. The planting was taken up in March 2002 when the water level in the tank as well as the canal was full. Among the species planted, survival percentage and biomass production of *Eucalyptus tereticornis* was higher compared to banana. The survival percentage of *Eucalyptus tereticornis* was 94 per cent whereas in banana it was 65 per cent. The growth performance of 18 months old eucalyptus in bio-drainage area was better compared to the eucalyptus trees planted in adjacent areas. At the end of first year, leaves, pseudo stem, flowers, and fruits were harvested from the banana crop. *Eucalyptus tereticornis* can be harvested at the end of fourth year and it can be allowed for coppicing. By coppicing there will be more number of shoots and foliage emanating from the present single stump thus resulting in increased evapotranspiration and biomass. After the introduction of bio-drainage, water stagnation, foul smell, mosquito problem and harmful weed growth were not observed. As there is no stagnation of water, problem of dampness in the nearby fields is reduced and hence the cultivation could be taken as planned.

From this pilot study, it can be concluded that it is very economical and environment friendly to introduce bio-drainage system for the degraded drainage problem lands in low lying delta areas. It is quite common in delta areas that the annual crops like paddy, pulses etc., are washed away during monsoon floods. However, the tree crops used for bio-drainage, can withstand the inundation, quickly drain the water, protect the annual crop nearby and also give revenue in terms of biomass production. Thus, by adopting the low-cost bio-drainage system, the farmers in the delta region can improve their crop productivity.

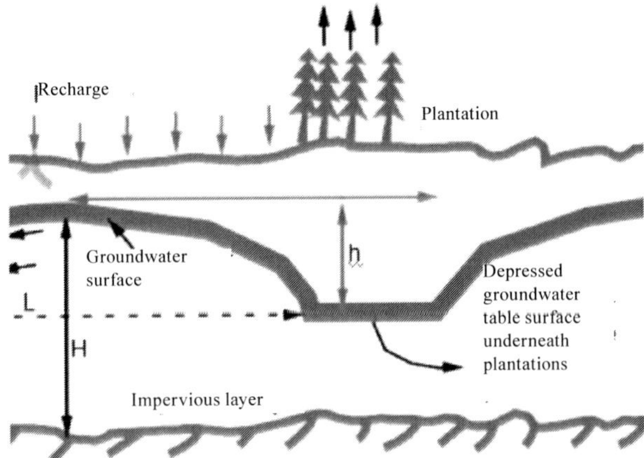

Figure 1: Plantation effect on Biodrainage

Plate: 1 Study area with Eucalyptus tereticornis

IV - Eucalyptus in wetland agroforestry: A case study of ecological benefits and site productivity.

A study was conducted at Agricultural Engineering College and Research Institute, Kumulur, Tiruchirappalli District Tamil Nadu (100 4' N, 780 3'E 70 m above sea level) to find out the effect of *Eucalyptus tereticornis* interpolating with rice crop compared with rice mono culture on ecological benefits and site productivity. Two varieties of rice crop var. Co 51 & Co 52 were inter plated (Plate 2 & 3) in the 10 years old *Eucalyptus tereticornis* plantation with an espacemant of 3m × 1.35 m compared with the same rice verities planted in mono culture with same environmental and edaphic condition. Soil samples were collected before transplanting and after harvesting of rice crop, and analyzed for organic carbon (%), available nitrogen, phosphorus and potassium (kg ha^{-1}) and compared with open field condition (Table 4). Tree survival (%) and tree girth (cm) of eucalyptus were measured before and after harvesting of rice crop. During the growth period of rice crop, the growth parameters viz., plant height (cm), leaf length (cm), leaf width (cm), no. of leaves, No. of tillers, no. of productive tillers and dry matter production were recorded in different stages. The allelopalhic studies were conducted using the soil samples collected from the plantation and open field condition with the seeds of groundnut, black gram, green gram, maize and rice and compared with control. The yield of CO51 and CO52 in Eucalyptus inter planting system observed was 840 kg/acre and 1040 kg/acre respectively and in rice mono culture system it was 1850 and 2000 kg/acre respectively. The results indicated that the profits of the inter plantations (rice+ wood biomass) per unit area is more than that of rice mono culture. Hence, Eucalypts agro forestry is a profitable option for farmers, in form of economically viable, environmentally sound and technically workable in wetland plain areas.

Plate 2 Eucalyptus + Rice intercropping (Milky stage)

Plate 3 Eucalyptus + Rice intercropping (Physiological Maturity)

Table 4: Organic carbon (%) and NPK (kg ha^{-1}) content in Soil under Eucalyptus + Paddy system and in open field.

Soil Analysis	Eucalyptus + Paddy system				Open field			
	OC	N	P	K	OC	N	P	K
Initial	0.54	181	11.5	126	0.44	173	11.0	115
Max tillering	0.57	183	11.5	128	0.45	176	11.1	121
Post harvest	0.57	188	11.7	130	0.44	177	11.7	130

Table 5: The grain yield of rice (Var. CO51 and CO52) monoculture and Eucalyptus inter planting system.

Rice varieties	Grain yield (Kg/ac.)	
	Rice monoculture	Eucalyptus inter planting system
CO51	1850	840
CO52	2000	1040

V- Other Identified Agroforestry systems for coastal zone

i. Coastal Home gardens
ii. Coastal vegetation
iii. Silvopastoral systems (Grazing under plantations)
iv. Fodder farming on neglected coconut plantations and other areas
v. Multiple Purpose Trees (MPTs) and shrubs on farm lands,
vi. Alley cropping
vii. silvihorticulture
viii. Aquaforestry

i. Costal Home gardens

(a) Zone of Distribution: Along The Malabar coast and Andaman islands
(b) Tree component: Coconut, Arecanut, Jack, Mango, Citrus etc.
(c) Annuals: Tomato, Brinjal, and Chilli.
(d) Poultry: hen, cock and duck.

Services provided by home gardens are: (a) food and nutritional security, (b) habitat protection, (c) soil and water conservation, (d) environmental services and (e) high rate of carbon sequestration

ii. Coastal Vegetation

A recurring pattern of dune, cliff top, mangrove and salt marsh plant communities characterizes many rocky coasts. The plant communities comprise a diverse variety of plant species representing a number of plant families. In general, the plants are adapted to cope with the hot, dry conditions and saline environments. The communities form important habitat for a variety of native fauna. They also have a very important role stabilizing and trapping marine sediments and forming a protective buffer between the land and the sea.

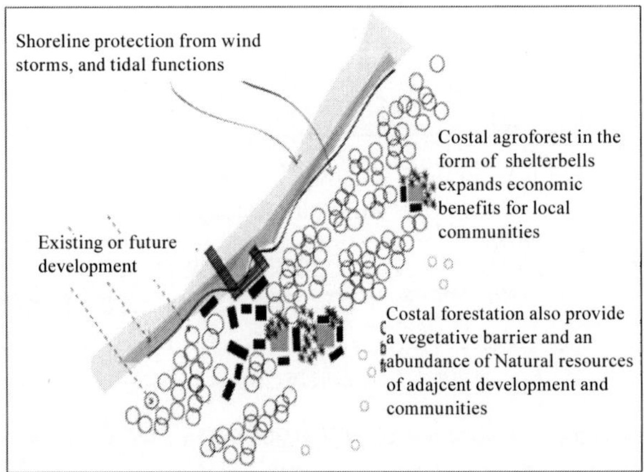

iii. Silvo-Pastoral Systems (Cut and Carry Systems/Fodder Farming)

The fodder species such as Pennisetum purpureum, Panicum maximum, Brachiaria ruziziensis and Euchlaena mexicana grown in association with Casuarina equisetifolia and Ailanthus malabarica.

- Forage grasses such as *Coix lachryma-jobi, Brachiaria mutica, and Echinochloa spp.* can successfully be cultivated
- Multipurpose trees such as *Aegle marmelos, Artocarpus spp., Bauhinia variegata, Erythrina variegata, Grewia glabra, Hibiscus tiliaceous, Moringa oleifera, Pitchecelobium dulce, Pongamia pinnata, Samanaea saman, Sesbania grandiflora and Trema tomentosa* can also be successfully incorporated.

iv. Fodder farming on neglected coconut plantations and other areas

Live Fences and Hedges

- Many trees are found grown on field boundaries, which are used as multipurpose trees by the farmers.
- *Acacia nilotica, Ailanthus excelsa, Bambusa spp., Borassus flabellifer, Casuarina equisetifolia, Cocos nucifera, Carissa carandas, Cordia rothii, Dalbergia sissoo, Ficus spp., Leucaena leucocephala, Moringa oleifera, Prosopis juliflora, Syzygium cuminii, Tamarindus indica and Ziziphus mauritiana* are very frequently found on bunds or farm boundaries

v. MPTs in coastal regions

1.	Plantation crop combination, multistoried	Production of multiple outputs, cash	Composite fish culture in ponds and multipurpose trees in homesteads
2.	Trees with aquaculture	Fish, fuel, fodder, timber.	*Acanthus ilicifolius, Avicennia officinalis, Carbera odollam, Rhizophora conjugata etc*
3.	Mangroves plantations as part of homesteads.	Shore protection, fuel, fodder, environmental protection	*Azadirachta indica, Casuarina equisetifolia, Prosopis chilensis, Acacia senegal etc.*
4.	Shelter belts and wind breaks	Shore/beach stabilization	*Aegle marmelos, Albizia spp., Azadirachta indica, Bamboos, Bombax malabaricum, Calliandra calothyrsus, Cassia spp. etc.*
5.	Trees on boundaries of agricultural fields (Agrisilviculture)	Fodder, fuel, shade, minor products	*Aegle marmelos, Albizia spp., Azadirachta indica, Bamboos, Bombax malabaricum, Calliandra calothyrsus, Cassia spp. etc.*

Tree species used in coastal plantations

1. *Casuarina equisetifolia*
2. *Terminalia catappa*
3. *Lannea coromendalica*
4. *Tamarix sp.*
5. *Cocos nucifera*
6. *Glochidion sp.*
7. *Calophyllum inophyllum*
8. *Barringtonia asiatica*
9. *Palmae metroxylon sagu*
10. *Nypa fruticans*
11. *Pandanus sp. [palm tree]*
12. *Scaevola sericea [shrub]*

vi. Hedge Row (Alley) Cropping

These practices are usually adopted for sloping lands where forage shrubs are planted across the slope and forage grasses and legumes or crops in the interspaces.

Suitable species: *Gliricidia sepium, Leucaena leucocephala, Cassia siamea, Morus alba, and Pithecelobium dulce.*

vii. Silvi-Horticulture

This system is defined as growing of trees and fruit trees or ornamental trees or vegetables/flower together in same lands at the same time. This system is common in home gardens of mid-hills, Terai and Inner Terai of Nepal, where fodder trees such as Badahar, Tanki, Ipil Ipil etc., and timber and fire wood species such as Sissoo, Eucalyptus, Baikaino, etc., are grown around fruit orchard that act as shelter belt, and horticultural crops such as ginger, turmeric, yam, colocassia and vegetables are grown under fruit trees.

The main advantages of this system are as follows:

(a) Produce multiple products such as food, fruits, fodder and forage needed for livestock, fuel wood, timber, and leaf litter needed for organic manure production.
(b) Improve and sustain the livelihoods of farmers by increasing the level of income through the sale of fruit/vegetables.
(c) Trees grown around fruit gardens also provide extra-income.
(d) This is also the best practice for soil nutrient recycling, which also helps to reduce chemical fertilizer purchase.
(e) Improve the farm site ecology by reducing soil erosion and nutrient loss.
(f) Improve the local micro-climate and enhance the productive capacity of the farm.
(g) This practices helps for the beautification of the surrounding areas.

Successful combinations in coastal stretch

The following tree species function successfully within a shelter belt and/ or common coastal and lowland species that will contribute to the re-establishment of lowland forest systems

Trees

(a) Rubber Tree - *Ficus elastica*
(b) Sumatran Pine - *Pinus merkusii*
(c) Coconut tree - *Cocos nucifera*
(d) Acacia - *Acacia mangium*

Intercrops

(a) Phase I and II : Sweet potato, Soybean, Cinnamon, Rice, Leguminous crops, Groundnut
(b) Phase III: Maize, cacoa, Banana, nutmeg, clove, Pineapple

Species Composition

(a) Species selection determines the success of shelter belts.

(b) It is important to choose native tree species of appropriate hardiness, foliage, uniform canopy, and good branch retention for taller species that occupy the center (windbreak) of the shelter belt.

(c) Smaller trees and shrub species should be planted on the outer edges of a belt.

Intercropping

- Intercropping techniques can be implemented between tree rows to encourage agro forestry practices within a shelterbelt while supplementing agricultural revenue.

viii. Aquaforestry

(a) Shrimp or fish farming is one of the viable commercial, alternatives to agriculture coastal areas.

(b) The development of brackish water aquaculture especially shrimp farming has been found to have substantial economic giants.

(c) Aquaculture keeping mangroves intact is most feasible and sustainable option for promotion of aquaculture in inundated areas.

Suggestion to improve the status of agroforestry research in coastal areas (Dagar, 2000)

(a) A time-frame programme must be chalked-out for rehabilitation of mangroves along coasts. Techniques are available for planting of mangroves, a mass programme should be framed and implemented without further waste of time.

(b) Aquaculture research keeping mangroves intact should to given priority.

(c) Genetic improvement of identified potential multi- purpose trees such as palmyrah palm, tamarind, nipa palm, arecanut, coastal badam, pandanus, kapok, etc.

(d) Attention must be paid to organic farming, mycorrhizae in relation to agroforestry and integrated pest-management in agroforestry.

(e) Developing the network for transferring the proven technology to the farmers.

(f) Developing mathematical and statistical models after evaluating the successful agroforestry systems/practices.

(g) Exploring unexploded and under-exploited plants of high economic value such as medicinal aromatic and oil-yielding plants.

(h) Research efforts should be made to explore marine algae or seaweeds for food, medicine and green manuring.

(i) Most emphasis should be given in evolving viable techniques for rehabilitation of degraded salt-affected areas involving good quality forages, MPTs, plantation crops and plants of industrial application.

(j) Research on quality assessment of the products obtained in saline habitats must be given priority.

(k) Promotion of national and inter-national collaborative research programmes.

(l) Research on saline agriculture including irrigation with saline water, developing halophytic crops, involving fish, shrimp culture and poultry with agroforestry systems, genetic improvement of salt- tolerant plants, raising nursery with saline water and multiplication and conservation of useful genetic material should be given immediate attention.

(m) Developing more agro-based industries in coastal areas to create more employment through value addition to the product based on coconut, oil palm, honey, cashew nut, rubber, fruit, fish & shrimp, milk, beverages, poultry, sea food and mangrove products.

Conclusion

Coastal zone in India is a crucial ecosystem which are currently facing threats due to over exploitation and also due to climate change scenarios. On the other hand, this ecosystem supports livelihood of a major population in the country. The three major problems in the coastal zone are heavy wind, salt affected soils and water logging. There are trees based solution to address these three major issues viz. Windbreaks and Shelter belts to protect the crops from wind damage, deploying of salt tolerant tree species to reclaim salt affected soils and Biodrainage systems using trees to recover the water logged fields. With regard to Windbreaks, it is important to note that Institute of Forest Genetics and Tree Breeding, Coimbatore has released Windbreak Tree Varieties in Casuarina to establish a very windbreak systems. With regard to salt affected soils, CSSRI, Karnal has identified suitable tree species for various types of salt affected soils and also technologies to reclaim the salt affected soils. Tamil Nadu Agriculture University has developed technology package and identified suitable tree species to address the water logging problems in the coastal zone. Thus, by adopting the identified potential tree varieties/species and the suitable trees based technologies, there exists tremendous potential for sustainable livelihoods for the peoples inhabited in the coastal ecosystem.

References

Abrol, I.P. and Bhumbla, D.R. (1978). Some comments on terminology relating to salt affected soils. Proc. Meetings Sub-Commission Salt affected soils, Dry land Saline Seep Control, 11th Int.Soil Sci., Congr. Edmonton. pp 6–27.

Abrol,I.P. and Sandhu,S.S. (1980).Growing trees in alkali soils. *Indian Farming.* **30:** 19–20

Ali, J. D., Singh, A. and Matthew, W.(2018). Sustainable coastal ecosystem management – An evolving paradigm and its application to Caribbean SIDS. Ocean & Coastal management, **163:** 173–184.

Buvaneswaran, C., Vinoth Kumar,K., Velumani, R.and Senthilkumar, S. (2018). Experimental Design for Evaluation of Clones of Casuarina for Windbreak Agroforestry System. *Journal of Tree Sciences,* **37 (1):** 11–18.

Campi, P., Palumbo, A.D. and Mastrorilli, M. (2009). Effects of tree windbreak on microclimate and wheat productivity in a Mediterranean environment. European *Journal of Agronomy*, **30(3):** 220–227.

Dagar, J.C. (2000). Agroforestry Systems for Coastal and Island Regions. *Indian Journal of Agroforestry.* 2:59–74

Dhyani, S. K. (2018). Agroforestry in Indian Perspective. In: Rajeshwar Rao G, Prabhakar M., Venkatesh, G., Srinivas, I. and Sammi Reddy, K. (Eds.). (2018). Agroforestry Opportunities for Enhancing Resilience to Climate Change in Rainfed Areas, ICAR - Central Research Institute for Dry land Agriculture, Hyderabad, India. p. 224

Dwivedi, A.P. (1992). Agroforestry: Principles and Practices. Oxford and IBH Publication Co., New Delhi.

Gill,H.S., Abrol, I.P. and Gupta, R.K. (1990). Afforestation of Salt-affected soils. Soils. CSSRI, Karnal. pp 1–6.

Gupta, J.P., Kar, A.and Faroda,A.S.(1997). Desertification in India: problems and possible solutions. *Yojna,* **41**: 55–59.

IDF.(1981). *La réalisation pratique des haies brise-vent et bandes boisées.*Institut pour le Développement Forestier, aris, 129 pp. cited from 'Trees outside forests -Towards a better awareness'. FAO CONSERVATION GIUDE No. 35.

Marale, S.M. (2013). Strategies for coastal ecosystem management in India. *Environment, Development and Sustainability,* **15(1):** 23–38.

Masilamani, P. and Santhana Bosu, S. and Annadurai, K. (2003). Bio drainage for degraded drainage problem lands. *Leisa India*, **5**:19.

Szaboles, I. (1980). Saline and Alkali soils- Commonalities. Pro, Int, Symp. Salt Affected Technologies for wasteland development (Eds. Abrol, I.P and Dhruva Narayana, V.C). ICAR Publications, New Delhi-12, 335–380 pp.

Tyndall, J. C., Wallace, D. C. (2011). Windbreaks: a "fresh" tool to mitigate odors from livestock production facilities. *Agroforestry Notes*, **41**: 4.

Yadav, J.S.P. (1980). Salt affected soils and their afforestation. *Indian Forester.* **106:** 259–272.

15

Solid Waste Management and Environmental Awareness The Need of the Hour

Vasanthy Muthunarayanan
Assistant Professor, Water and Solid Waste Processing Laboratory
Dept. of Environmental Biotechnology, School of Environmental sciences,
Bharathidasan University, Tiruchirappalli-620 024. Tamil Nadu, India.

Introduction

One of the major problems being faced by cities and towns relate to management of municipal solid waste (MSW). Waste quantities are increasing day by day due to the increasing population and the municipal authorities are not able to upgrade or scale up the facilities required for proper management of such wastes. In many cities and towns, garbage is littered on roads and foot-paths. Citizens are also not accustomed to use the available storage facilities (dust bins) set up by the authorities. At large, lack of organized system of house-to-house collection of waste has created the littering habits. By and large, hardly we can see any city/town complying with the Municipal Solid Wastes (Management and Handling) Rules, 2000 in 'totality". As per information provided by Central Pollution Control Board (CPCB), 1,27,486 TPD (Tons per day) municipal solid waste was generated in the country during 2011–12. Out of which, 89,334 TPD (70%) of MSW is collected and 15,881 TPD (12.45%) is processed or treated.

In spite of various measures taken towards the proper waste collection, segregation and disposal, proper waste processing units were established as per the following details as per CPCB.The waste processing technologies reported in the country are; composting, vermi composting, biogas plant, RDF-pelletisation and others. Some of the sepelletization plants are associated with power plants for generation of electricity. However, mechanical composting and vermi composting are more popular in the country.

Solid Waste Disposal

The normally preferred and widely followed methods of solid waste disposal include (1) Open dumping and (2) Landfills.

1. Open dumping

Open dumping occurs when large quantities or piles of waste accumulate in areas not designed to handle such materials. Open dumps are commonly found in forests, backyards, abandoned buildings and swimming pools. Open dumps are usually removed shortly after they are created. Open dumps are usually formed when making a foundation for a building. Even though the waste of an open dump would decay and form part of the soil as organic manure, they would also breed pests and vectors which cause diseases to those living near an open dump.

Such practise of open dumping is found to pose certain hazards as follows:

Physical hazards– Open dumping often presents physical hazards with broken glass, sharp metal on discarded items, and appliances in which children or animals can be trapped.

Chemical hazards– Disposed chemicals may be toxic to a child who goes onto the site. Household hazardous waste (HHW) such as paint, pesticides and other toxic chemicals, can be found in open dumps.

Biological hazards– Items such as syringes or other discarded medical items,household garbage, which may include food scraps and dirty diapers, attracts pathogens.

The chemicals and other contaminants found in solid waste can seep into our groundwater and can also be carried by rainwater to rivers and lakes that provide essential wildlife habitat. These contaminates can also end up in our ground water, rivers and lakes that are our sources for drinking water.

Health risks associated with open dumping

The health risks associated with illegal dumping are significant. Areas used for open dumping may be easily accessible to people, especially children, who are vulnerable to the physical (protruding nails or sharp edges) and chemical (harmful fluids or dust) hazards posed by wastes. Rodents, insects, and other vermin attracted to open dump sites may also pose health risks. Dump sites with scrap tires provide an ideal breeding ground for mosquitoes, which can multiply 100 times faster than normal in the warm stagnant water standing in scrap tire causing several illnesses. Poisoning and chemical burns resulting from contact with small amounts of hazardous, chemical waste may get mixed with general waste during collection & transportation.

2. Sanitary landfill

Sanitary Landfills are designed to greatly reduce or eliminate the risks that waste disposal may pose to the public health and environmental quality. They are usually placed in areas where land features act as natural buffers between the landfill and the environment. For example, the area may be comprised of clay soil which is fairly impermeable due to its tightly packed particles, or the area may be characterised by a low water table and an absence of surface water bodies thus preventing the threat of water contamination. In addition to the strategic placement of the landfill, other protective measures are incorporated into its design.

The bottom and sides of landfills are lined with layers of clay or plastic to keep the liquid waste, known as leachate, from escaping into the soil. The leachate is collected and pumped to the surface for treatment. Boreholes or monitoring wells are dug in the vicinity of the landfill to monitor groundwater quality.A landfill is divided into a series of individual cells and only a few cells of the site are filled with trash at any one time. This minimizes exposure to wind and rain. The daily waste is spread and compacted to reduce the volume, a cover is then applied to reduce odours and keep out pests. When the landfill has reached its capacity, it is capped with an impermeable seal which is typically composed of clay soil. Some sanitary landfills are used to recover energy. The natural anaerobic decomposition of the waste in the landfill produces landfill gases which include carbon dioxide, methane and traces of other gases.

These landfills present the least environmental and health risk and the records kept can be a good source of information for future use in waste management. However, the cost of establishing these sanitary landfills are high when compared to the other land disposal methods.

Health risks associated with Landfills

Gases escaping from landfills contain toxic pollutants that can cause cancer, asthma, and other serious health effects. These gases were reported to carry toxic chemicals such as paint thinner, solvents, pesticides, and other hazardous volatile organic compounds. All dumps also leak toxic leachate; even "state-of-the-art" landfills will eventually leak and pollute nearby groundwater.

Landfills are also a significant contributor to climate change. They are the largest global source of human-created methane emissions, a toxic climate-changing gas that is 25 to 72 times more potent than carbon dioxide.

Aerobic and anaerobic systems for solid waste treatment: The treatment of the organic solid wastes can be carried out either in the presence or absence of oxygen. Three common methods of processing include composting, vermicomposting, and bio-methanation/anaerobic digestion, among which vermicomposting is found cheaper,effective and an eco-friendly.

Vermicomposting is the process in which worms are used to convert organic material (usually waste) into humus like material known as vermi compost. The goal is to process the material as quickly and efficiently as possible. The earth worms maintain aerobic conditions in the vermi composting process, ingest solids and convert a portion of it to earthworm biomass and respiration products and digest peat like material termed as worm castings. Castings are much more fragmented, porous and microbially active than parent material due to humification and increase decomposition. These earthworms derive their nourishment from microorganisms involved in waste decomposition and organic waste to be decomposed. During this process, important plant nutrients such as nitrogen, potassium, phosphrous, calcium etc. present in the waste are converted through microbial action into form that are much more soluble and available to plant than those in parent substrate. Overall, the vermi composting process is a result of combined action of earthworms and microflora living in earthworm intestine and in the organic waste. It is an aerobic, biooxidation and stabilization, non thermophilic process of organic waste decomposition that depends upon earthworm to fragment, mix and promote microbial activity. The basic requirements during the process of vermicomposting include suitable

- Bedding
- Food source
- Adequate moisture
- Aeration,
- Suitable temperature
- pH

Post consumer wastes

Post-consumer waste is a waste type produced by the final consumer, which cannot be recycled. These are biodegradable and are extracted from renewable source. As per the data given by CPCB2012, the per capita generation 560 grams/day. Out of various post consumer wastes, paper cups are one of the waste disposed in many of the petty shops, educational institutions, marriage halls, public and private offices etc. Hence a study was taken up and he paper cup wastes were subjected to for vermi composting process with the earthworm namely, *Eudrilus eugeniae* (African night crawler).

The research work highlights the feasibility of managing the generated paper cup waste in an eco-friendly manner. The physico-chemical analysis of the vermi compost prepared using paper cup waste revealed that the Type A (1:1) i.e., equal composition of paper cup and cow dung was found to yield a compost of manural value. Nine different species of cellulose degrading bacteria were isolated and characterized from the vermicompost. Consortia of cellulose degrading bacteria have the potentiality to serve as an inoculum for enhancing the degradation of paper cup waste. Thereby the overall period of degradation was reduced from 6 to 3 months. It was found that earthworm has a unique nature of separating the plastic coating in the paper cups.The morphological and physiological changes in earthworms and its regaining capacity were understood. Hence the vermicomposting technology can be an effective, efficient and an eco-friendly technique to manage the degradable waste.

To conclude. management of solid waste is the most essential step to safeguard the environment and to ensure good health. With a per capita generation of 500-600 g/day, segregation of such wastes is a must for its proper management. As a contribution by every house, the domestic degradable waste could be managed by vermicomposting to reduce the overall bulk of the municipal solid waste.

Environmental awareness - the need of the hour

Environment includes all living organisms and non-living objects. The environmental resources like air, land and water facilitate us to lead our life and to meet our needs. Development also means meeting the needs of the people. But the population explosion, has made us to pressurize the environment to meet our increasing demands. When the pressure exceeds the carrying capacity of the environment to repair or replace itself, it creates a serious problem of environmental degradation. If we use any environmental resource such as ground water beyond its limit of replacement, we may lose it forever. Therefore, there is a need to create 'awareness' about Environmental protection. While efforts are being made at the national and international level to protect our environment, it is also the responsibility of every citizen to use our environmental resources with care and protect them from degradation.

Anthropogenic activities

All human activities have an impact on environment. But in the last two centuries or so, the human influence on environment has increased manifold due to the rapid population growth, urbanization and industrialization. The anthropogenic activities have resulted in the pollution of land and water thereby affecting the plants, animals and human beings. The water quality of the surface and ground water is affected due to the discharge of domestic sewage, agricultural wastes and industrial effluents. This has affected the natural regeneration capacity of the aquatic ecosystems. The continuous discharge of effluents has now affected the water quality parameters and has added heavy metals, pesticides, fertilizers, detergents, and so on into the aquatic ecosystem posing threat to the existence of the aquatic organisms.

With billions of people on the planet earth, sewage disposal is a major issue for many of the developing countries. As per WHO (2016), about 663 million people don't have access to safe drinking water and about 2.4 billion people don't have proper sanitation. Due to the contamination of sewage with drinking water supplies, the break out of water borne diseases happen regularly and endemic diseases break out at regular intervals. We need to next think about the nutrients such as nitrate and phosphate added to the aquatic systems which have facilitated the growth of algal blooms due to the eutrophicated condition of the water system. This condition may even result in ecological succession too, thereby converting the aquatic ecosystem to terrestrial ecosystem. Especially the marine ecosystems are now threatened by the plastic debris along with the oil spill too. These plastics were found to the reason for the death of many numbers of aquatic organisms.

Soil pollution

The quality of soil is deteriorating resulting in the loss of agricultural land. The loss is estimated to be about five to seven million hectares of land each year. Soil erosion, as a result of wind and/or water, damages the environment. The recurring floods have their own peculiar casualties like deforestation, silt in the river bed, inadequate and improper drainage, loss of men and property. The vast oceans, after being turned in to dumping grounds for all nuclear wastes, oil spills have poisoned and polluted the whole natural environment.

Air pollution

The air pollution also has two sources, namely the natural and anthropogenic i.e., the varied sources include burning of fossil fuels for domestic purposes and transportation, industrial emissions etc. which has resulted in the emission of greenhouse gases in the atmosphere, thereby making a way for the climate change and global warming too. The ozone layer depletion in the stratosphere, occurrence of acid rain, formation of photochemical smog now and then are all results of such anthropogenic activities. The inhaling of these air pollutants pose health hazard to the living organisms and also damages the non living objects too.

The segments of the environment were already polluted namely, the hydrosphere, the lithosphere and the stratosphere thereby the biosphere to a maximum. To mention a few, the following pollution episodes would make us remember what we have done to our environment. Such pollution episodes include:

- Breakdown of minamata disease (1956)
- London smog (1972)
- Ozone layer depletion
- Love canal episode (1970)
- Biodiversity loss
- Eutrophicated aquatic systems
- Loss of forest coverage
- Acid rain affecting Taj Mahal

Sustainable Development

Under these circumstances, theWorld Commission on Environment and Development (the Brundtland Commission) submitted its report entitled "Our common future' in 1987. which highlighted and popularised the concept of 'Sustainable Development'. Sustainable development has been defined on meeting the needs of the present generation without compromising the need of future generations. The aim should be to achieve sustainable levels of people's welfare and development.

Creating awareness

To achieve such sustainable development, it is essential to create awareness regarding the environmental issues to the common public. The four levels of awareness include

- **First**: It is realizing the symptoms of Environmental deterioration.
- **Second**: One should understand that the reason for environmental pollution includes not only industries but also the ever increasing population.
- **Third**: We should realize that protecting and preserving the environment alone could save mother earth.
- **Fourth**: One should understand the following doctrines to ensure sustainable living to all the organisms.
- ▪ All species are interconnected.
- ▪ The work of human is not to rule nature, but to work in par with nature.
- ▪ Backfire of the mechanisms may happen at any time, if we overrule nature.

- One should set the goal to preserve the biological along with social integrity, sustainability and diversity of life supporting systems.

Steps to be taken

To facilitate and to create such awareness the following steps could be taken:

- The children of kindergarten to students of secondary grade must be educated regarding the Environmental issues.
- Practical oriented education must be stressed.
- The values of the four segments of the environment must be understood by every citizen.
- Urban and rural public must be educated through short films, tv ads, banners, processions, public meetings and through mass media.
- Prizes and concessions could be given to the industries following the norms.
- Usage of renewable energy resources could be encouraged.
- NGO's could be encouraged to work in this area.
- Plantation could be taken as foremost duty of every individual.
- Let us save our earth as it is the only planet we are blessed with.

To ensure sustainability, the following steps could be practiced:

- Reduce the waste of matter.
- Emphasize pollution prevention.
- Reduce, Recycle,Reuse and Rethink.
- Make goods that lasts Longer.
- Rely on renewable Energy Resources.
- Sustain earth's biodiversity.
- Use non renewable energy resources reasonably.
- Discourage earth degrading behavior.
- Reduce poverty by hard work.
- Aim at Sustainable development.

To conclude, it is also true that the government is taking adequate steps for imparting the awareness to the public at different levels. This is the reason for creating Sarita Vihar in South Delhi into a zero-waste residential colony, preparing vermicompost at Chokhi Dhani (fine hamlet), a five-star ethnic village resort, using biodiesel for vehicles, replacing CFC with HCFC to reduce the ozone layer depletion, using unleaded petrol, installation of solar panels for energy harnessing, initiating Swachh Bharat Abhiyan mission to make Clean

India is all giving the understanding that many measures are being taken to ensure sustainability. But still it is high time to raise the consciousness of common public to clearly understand that this is the only planet we are left with and hence our efforts to be a maximum to restore its pristinity.